INTERCORTICAL SYSTEMS OF
THE HUMAN CEREBRUM

INTERCORTICAL SYSTEMS
OF THE
HUMAN CEREBRUM

MAPPED BY MEANS OF NEW
ANATOMIC METHODS

BY

JOSHUA ROSETT

ASSISTANT PROFESSOR OF NEUROLOGY, COLUMBIA UNIVERSITY
SCIENTIFIC DIRECTOR, BRAIN RESEARCH FOUNDATION

NEW YORK   M·CM·XXXIII
COLUMBIA UNIVERSITY PRESS

COPYRIGHT 1933
COLUMBIA UNIVERSITY PRESS
———
PUBLISHED 1933

PRINTED IN THE UNITED STATES OF AMERICA
WAVERLY PRESS, BALTIMORE, MARYLAND

TO
OLIVER S. STRONG
TEACHER, COLLEAGUE AND FRIEND
THIS BOOK IS DEDICATED

FOREWORD

Anyone who has worked on the cerebral cortex and the cerebral hemisphere, with its intricate structure, must have been keenly interested in the ingenious method devised by Dr. Rosett to decompose the tissue into layers that can be studied in greater detail and more in harmony with the course of fibers than is possible with the ordinary serial sections. With the additional improvement of using a process of explosion and what we might call autoanatomy, Dr. Rosett has succeeded in getting a splitting up along lines of least resistance, far superior to manual dissection. The greatest obstacle to study and utilization of similar material has always been the difficulty of resynthesis. With the highly skilled and extraordinarily well systematized plan of making casts of the complete brain and its subdivisions, open to the view of the worker on the sections, Dr. Rosett has succeeded in facilitating the allocation of the subdivisions in an unprecedented manner. No matter how useful the degeneration methods and especially the Marchi method may be, the final checking up with direct histological control becomes an evident postulate to any investigator. It is very gratifying that the first orienting study in this field becomes available in a form allowing the investigator a direct study of the detailed observations and the possibility of orienting for himself. We are here dealing with a life work; through the presentation of this first contribution, it will become possible to recognize opportunities for further research and for utilization of results already obtained.

The material outlined in this book naturally derives its chief orientative importance from the fact that it is a unique collection of observations which will be added to and expanded to include the detailed study of the special regions and the functional mechanisms. Its correlation with the more superficial results reached by easier methods will be the goal of monographic presentations, for which the present volume presents the method and the inspiration.

ADOLF MEYER,
Professor of Psychiatry
Johns Hopkins Hospital and University

BALTIMORE
September, 1933

PREFACE

I have long admired the ingenuity and versatility of Dr. Joshua Rosett in applying new methods to biological problems. It was with great interest, therefore, that I first heard of his technique for mapping out the association fiber systems of the human cortex. He described his plan of work at a meeting of the Program Committee of the Association for Research in Nervous and Mental Disease in December, 1929. The committee, of which I was chairman, approved it, and urged Dr. Rosett to continue work and report progress at the next meeting of the Association in December, 1930, when the subject of discussion was to be "The Convulsive State." Obviously such an extensive and meticulous research could not be hurried, so it was at a meeting of the American Neurological Association in May, 1931, that the first (preliminary) report was read, followed a year later by a second report. The following pages present in detail a great deal of investigative work, but, as the author modestly states, this is only a start that must be expanded and corroborated.

At first glance it might seem a long step from anatomical studies of intercortical connections to the cause of convulsions. Most investigators agree, however, that epileptic convulsions usually originate in some part of the cerebral cortex, and spread rapidly to involve large portions of the brain. This spread is an important part of the convulsion, since it determines the way in which the abnormal discharge will reveal itself—whether as muscular jerkings due to involvement of the precentral cortex, as paresthesia due to involvement of the postcentral cortex, or as scotoma due to involvement of the occipital cortex—to mention three of the many possibilities. The intercortical fiber systems are the pathways along which these epileptic discharges may spread in the cortical phase of the convulsion.

Apart from this special interest, however, it is of great importance to know accurately the anatomy of the cerebral cortex from all aspects. Only by understanding the fiber connections between gyri can the relationship of the recently discovered different areas of cortical cyto-architecture be understood, and upon these relationships depends much of the cerebral function. Upon a correct understanding of cerebral localization, depends our knowledge of how the brain functions. The brain is not an organ, it is a hundred organs rolled into one. The intricacies of its anatomy and physiology are as yet only schematically and fragmentarily understood. Upon the conviction that one day this form and function will be thoroughly known, rests our hope of at last understanding the biological basis of psychology; for no sound psychologist doubts that the brain is the organ of the mind. Dr. Rosett's work is an important step forward toward this goal.

STANLEY COBB,
Bullard Professor of Neuropathology
Harvard Medical School

BOSTON
September, 1933

ACKNOWLEDGMENTS

The costs of this work were defrayed by the Brain Research Foundation, by a grant of the Commonwealth Fund to the Neurological Institute for the study of epilepsy, and by the Department of Neurology, Columbia University.

The work has necessitated a large supply of post-mortem material, which was furnished by the State Infirmary and State Hospital for Mental Disease of Howard, R. I., through the kindness of the Chairman of the Rhode Island State Public Welfare Commission, Dr. Frederick Farnell.

I wish to thank Mrs. Louise Baker, Miss Mary Carter, Mrs. Ruth Kidder and Miss Betty Lowell for their valuable assistance.

J. R.

New York
September, 1933

CONTENTS

FOREWORD, by Dr. Adolf Meyer vii
PREFACE, by Dr. Stanley Cobb ix
ACKNOWLEDGMENTS xi

I. INTRODUCTION 1

II. METHOD AND TECHNIC 5
   1. Automatic Internal Dissection 5
   2. Manual Dissection, Plaster-of-Paris and Metal Casts 11
   3. Recording 14
   4. The Preparation of Flat Blocks 16
   5. Detailed Drawings 20
   6. The Preparation of Schemata from Detailed Drawings 21

III. THE SCOPE OF THE WORK 23

IV. REGARDING THE CELL ORIGIN OF THE INTERCORTICAL SYSTEMS . . 25

V. ANNECTANT CONVOLUTIONS 27

VI. INTERCORTICAL SYSTEMS OF THE MESIAL SURFACE OF THE CEREBRUM:
   A Note regarding the Method and Nomenclature Employed in
   the Enumeration of the Pathways 29
   1. The Paracingulate Fissure 29
   2. The Cingulate Fissure 33
   3. The Sulcus of the Cuneus 36
   4. The Calcarine and the Parieto-occipital Fissures 36
   5. Deep Intercortical Systems of the Occipital Lobe 45

VII. INTERCORTICAL SYSTEMS OF THE BASAL TEMPORO-OCCIPITAL AREA 52
   1. The Collateral Fissure 52
   2. The Inferior Longitudinal Fasciculus 54
   3. General Considerations regarding the Cingulate and Collateral
      Fissures 57

VIII. INTERCORTICAL SYSTEMS OF THE PARIETAL AREA 59
   1. The Interparietal Fissure 59
      The Inferior Postcentral Fissure—The Sagittal Ramus—The
      Transverse Occipital Fissure and the Sulcus Lunatus
   2. The Superior Segment of the Postcentral Fissure 66

CONTENTS

  3. The Posterior Ascending Limb of the Sylvian Fissure 67
  4. The Temporo-parietal Fissure 68
IX. INTERCORTICAL SYSTEMS OF THE CENTRAL AREA 70
  1. The Central Fissure 70
  2. Rare Types of the Central Fissure 74
X. INTERCORTICAL SYSTEMS OF THE FRONTAL AND PREFRONTAL AREAS . 84
  1. The Topography of the Frontal Lobe 84
  2. The Superior Precentral and the Superior Frontal Fissures . . 84
  3. The Inferior Precentral and the Inferior Frontal Fissures . . . 87
  4. The Middle Frontal Fissure (Eberstaller) 89
  5. Fronto-orbital Fissures 89
XI. DEEP INTERCORTICAL SYSTEMS OF THE LATERAL SURFACE OF THE
  CEREBRUM 91
  1. The Arcuate Fasciculus and the Intercortical Systems of the Insular Region 91
  2. General Considerations regarding the Arcuate Systems 102
XII. INTERCORTICAL SYSTEMS OF THE LATERAL TEMPORAL AREA 105
  1. The Superior Temporal Fissure 105
  2. The Middle Temporal Fissure 107
XIII. INTERCORTICAL SYSTEMS OF THE LATERAL OCCIPITAL AREA 109
  1. The Superior Lateral Occipital Fissure 109
  2. The Inferior Lateral Occipital Fissure 111
  3. The Temporo-occipital Fissure 113
XIV. GENERAL CONSIDERATIONS 114
  1. The Pattern of the Subcortical Pathways of the Fissures and Its Complications 114
  2. Criticism of Methods 116
  3. A Contribution to the Knowledge of the Causes of Fissuration . 117

REFERENCES 125

INDEX 129

ILLUSTRATIONS

Fig. 1A and 1B. Sagittal Sections from Different Portions of an Exploded Human Cerebrum Treated with India Ink 7, 8
" 2. Horizontal Sections from Different Portions of an Exploded Human Cerebrum Treated with India Ink 9
" 3. Coronal Sections from Different Portions of an Exploded Human Cerebrum Treated with India Ink 10
" 4. The White Laminae Underlying the Cortex of the Fissures of the Lateral Surface of the Left Hemisphere of a Human Cerebrum. 12, 13
" 5. The Next Step in the Dissection of the Preparation Shown in Fig. 4. 13
" 6. A Line Drawing of the Mesial Surface of the Plaster-of-Paris Cast of the Left Hemisphere. 15
" 7. A Boat or Basketlike Subcortical White Lamina of a Fissure . . . 17
" 8. A Diagrammatic Representation of the Several Steps in the Preparation of Flat Blocks. 19
" 9. A Photomicrograph (X 75) Showing Two Sets of Intercortical Fiber Bundles Crossing Each Other at Right Angles, in the Manner of a Textile Fabric, in the Subcortical Lamina of a Fissure 20
" 10. A Photomicrograph (X 35) Showing Curved Intercortical Pathways from a Subcortical Lamina of a Fissure 22
" 11. A Photomicrograph Showing the Arrangement of the Fibers in the Bundles and the Defects Produced by the Explosion of the Cerebrum 24
" 12. A Composite Schema of the Intercortical Systems of the Left Mesial Area of a Human Cerebrum 30
" 13. A Simpler Type of the Paracingulate Fissure than That Shown in Fig. 12 31
" 14. The Intercortical Systems of the Paracingulate Fissure Shown in Fig. 13 32
" 15. A Composite Schema of the Intercortical Systems of the Left Mesial Occipital Area of a Human Cerebrum. 37
" 16. Intercortical Systems of the Parieto-occipital and Calcarine Fissures of Three Human Hemispheres 43
" 17. The Gross and Microscopic Appearances of the Deep Intercortical Systems of the Occipital Lobe 48
" 18. The White Laminae Underlying the Cortex of the Fissures of the Mesial Surface of the Left Hemisphere of a Human Cerebrum . . 50
" 19. A Line Drawing of the Basal Surface of the Plaster-of-Paris Cast of the Left Hemisphere of the Cerebrum 53

ILLUSTRATIONS

Fig. 20. A Composite Schema of the Intercortical Systems of the Left Basal Area of a Human Cerebrum 55
" 21. A Line Drawing of the Lateral Surface of the Plaster-of-Paris Cast of the Left Hemisphere of the Cerebrum. 60
" 22. A Composite Schema of the Intercortical Systems of the Left Central and Parietal Regions of a Human Cerebrum. 61
" 23. The Central Fissure of a Right Hemisphere of the Cerebrum . . . 75
" 24. Detailed Drawings of Sections (Mags. 50 and 150) of Three Parts of the Subcortical Lamina of the Right Central Fissure Shown in Fig. 23, Illustrating the Method of Reconstruction 76
" 25. A Schema of the Intercortical Pathways of the Central Fissure of the Right Hemisphere of the Cerebrum Reconstructed from the Detailed Drawings in Fig. 24. 77
" 26. A Rare Type of the Left Central Fissure 78
" 27. The Intercortical Pathways of the Central Fissure Shown in Fig. 26 . 80
" 28. A Photomicrograph (X 130) of the Intercortical Hairpin Pathways from the Subcortical Lamina of the Left Central Fissure Shown in Fig. 27 80
" 29. A Rare Type of the Right Central Fissure 81
" 30. The Intercortical Pathways of the Fissure shown in Fig. 29 82
" 31. A Composite Schema of the Intercortical Systems of the Left Frontal and Prefrontal Areas of a Human Cerebrum. 86
" 32. Intercortical Systems of the Deep Fiber Sheet Lateral to the Intercortical Systems of the Fissures of the Insular Region 93
" 33A. The Insula and the Opercula of the Left Hemisphere Studied . . 94
" 33B. A Line Drawing of the Insula and the Opercula Shown in Fig. 33A . 94
" 34. A Composite Schema of the Intercortical Systems Underlying the Cortex of the Fissures of the Insula and the Opercula Shown in Fig. 33A and Fig. 33B 95
" 35. A Dissection of the Arcuate Intercortical Systems 97
" 36. An Arcuate Section of the Arcuate Fasciculus Shown in Fig. 35, Drawn at Mags. 50 and 150, Weigert-Pal Stain. 98
" 37. Parts of the Arcuate Intercortical Systems 99
" 38. A Dissection of the Fasciculus Centroparietalis 101
" 39. The Tangential Irregular Network of Single Nerve Fibers in the Lowest Cortical Layers of the Cerebrum of the Opossum . . . 103
" 40. A Composite Schema of the Intercortical Systems of the Left Lateral Temporal Area of a Human Cerebrum. 110
" 41. A Composite Schema of the Intercortical Systems of the Left Lateral Occipital Area of a Human Cerebrum 112

INTERCORTICAL SYSTEMS OF THE HUMAN CEREBRUM

CHAPTER ONE
INTRODUCTION

Beneath the cortex of the human cerebral fissures, and parallel to the surface, there extend vast and intricate systems of nerve fibers. Although the existence of these structures has been known to neurologists for over a century, the methods employed in mapping out their complicated network have been inadequate to the task. The conception of this network has therefore remained vague and largely erroneous.

As with most vague conceptions, the error in this instance has tended in the direction of over-simplification. Thus it has been the belief that the structures here dealt with consist of fibers which begin at points on the crest of a convolution, pass down the wall of the fissure, curve under its floor and rise on the adjoining wall to reach points on the crest of a neighboring convolution directly opposite that from which they started; hence the name U-fibers.

It will be shown later that such an arrangement is true only in part and in a limited sense. In none of the many microscopic preparations examined in this study could fibers be discovered which were long enough to extend from one convolution to another in any direction around the deeper fissures. Yet although the nerve fibers of these systems may be shorter than hitherto believed, many of the pathways in which they are grouped are certainly much longer. It will be shown that a large number of these pathways proceed diagonally as well as squarely across the subcortical laminae of the fissures; and that some of these diagonal pathways are as long or even longer than the fissures to which they may properly be said to belong.

The intercortical pathways which extend beneath the cortex of the fissures are to a large extent interrupted in their course in those protuberances of the cortex, in the depth of fissures, which are known as annectant convolutions. Notwithstanding the interruption of a number of these pathways, other contingents, in many instances, pass beneath the bases or around the annectants and they may thus persist in their course to the end of a given fissure or even beyond it. In one way or another, therefore, many of the pathways which underlie the cortex of the fissures establish connections between distant points.

It will appear in the present work that the subcortical white laminae of the fissures are by no means entirely constituted of the "short" intercortical systems. They will be shown to be intermixed to a certain extent with fibers of the longer intercortical systems, whose bulk is embedded deeper in the cerebrum, and with projection systems. In the microscopic sections of these laminae such extraneous fibers could not be clearly distinguished from the rest, but the fact that they contained extraneous elements was made manifest from the study of the anatomic relations of the deeper to the more superficial systems, and will be illustrated in the present report.

A better knowledge of these extensive intercortical systems calls for a change in the current points of view concerning the causes of a general and particular pattern of fissuration. It will appear that instead of being regarded as boundaries between structurally different areas of the cerebral cortex, the fissures are to be regarded as repositories of the greatest number of those subcortical tangential pathways which interconnect countless points in the cortex at different distances apart. Any particular pattern of fissuration must be, therefore, logically considered the effect rather than the cause of the particular arrangement of the underlying pathways.

The history of the preceding studies of the intercortical systems is repeated again and again in German works on the fiber systems of the cerebrum. A very thorough critical analysis of such studies may be found in Mayendorf's (1) monograph on the anatomy, physiology, etc. of the association systems. The nature of some of these studies will be gathered from the references made to them in the present report. For the rest they may be summed up in this place in a few lines. Arnold (2) appears to have been the first to call especial attention to the existence of the white lines underlying the gray substance of the tissues, not only of the cerebrum, but of the cerebellum as well. He gives, however, due credit to the anatomists who had mentioned these structures before him—to Reil, Rolando and Burdach. Meynert (3) and, following him, almost every investigator who has worked with and has written about the cerebrum, has dealt more or less with these structures in connection with his particular subject. Vulpius' (4) study of the tangential systems appears to be rather with reference to the lines of Baillarger and Gennari. In Poljak's (5) pictures of his degeneration experiments in animals there is a suggestion of degenerated fibers (pictured in red) in the wall of a sulcus adjoining the injured crest of a convolution. It is the clearest drawing of the kind I have encountered in the literature on the subject.

The early anatomists derived their ideas of the intercortical systems from the crude method of manual dissection. Most of the later anatomists, having discarded dissection, employed the less crude means of studying these systems from cross sections of cerebral tissue containing tracts of massive degeneration prepared by the method of Weigert-Pal. Still others, particularly Probst, to whose work reference will be made later, employed the excellent Marchi degeneration method in addition to that of Weigert-Pal. Flechsig, and a number of anatomists who followed him in the study of myelogenesis, agreed that that method was ill suited to the study of the course of the intercortical systems. All investigators have struggled with, and many have complained of, the difficulties and uncertainties inherent in any method which has for its basis the cross-sectioning of the tissue.

A number of the authors who have investigated the general subject of intercortical connections have taken the intercortical systems of the fissures for granted. Even so excellent an investigator as Quensel (6) does away with the entire subject by saying: "There is certainty regarding connections between two neighboring convolutions by means of short association fibers." Mayendorf's (1) opinions on the subject are varied and contradictory: He states plainly that the tracing of intercortical pathways in a series of cross sections from Weigert-Pal degeneration specimens is an impossibility. He rightly insists that the existence of given intercortical connections can be concluded with certainty only when the beginning and ending of the fibers may be plainly seen in a number of serial sections; and he asserts that "this is comparatively easy in the case of the U-bundles." But as he himself worked with cross sections of specimens of massive degenerations stained by the method of Weigert-Pal, it is impossible, in the light of the present study, to understand why it was easy for him to see the beginnings and endings of these fibers.

A large number of investigators merely mention the subcortical fiber systems of the fissures in connection with other subjects. None have attempted to map out these pathways in any one type of human or animal cerebrum.

INTRODUCTION

It is the latter task which has been the particular subject of this study. Yet although the results of the work to be described have taken a long time to obtain, it can by no means be said that the intercortical connections have been mapped out with any great degree of completeness. Even from the forthcoming report it will appear that the pathways in question differ in a number of details in different types of the human cerebrum; and there are a number of types. All that is hoped for from the present work, therefore, is to open a way for the discovery of a certain order in a realm of chaos.

The topography of the human cerebrum was studied for over a century by a large number of anatomists from many points of view. Reference to the work of some of those investigators will be made in the course of the present study. A review of the literature on this subject reveals an astonishing amount of ingenuity and painstaking labor devoted to the task of solving the mystery of the intricate and apparently hopelessly irregular pattern of convolutions wrought on the surface of the cerebrum. Many anatomists of our own generation, disdaining the study of the cerebral surface configuration, have been busily engaged in mapping out the extent of the cortex into a number of areas of different cell structure. So fruitful has been the latter line of research, that, as with every passing year the number of such cortical areas has increased, the suspicion has at last arisen that each separate gross irregularity of the cerebral surface is pregnant with physiological significance. One of the consequences of these neuroanatomical studies is that the question of the nerve connections between the different cortical cell areas is rapidly pressing to the front. Notwithstanding the necessity, however, no attempt was made in the present work to correlate the course and direction of any of the numerous intercortical pathways mapped out with any of the cortical cell areas delineated by any of the investigators of that subject. For it is the author's conviction that not until the present work has been repeated by many workers, its weak points, perhaps its errors, largely eliminated and many new data added, will a correlation of intercortical pathways and cell areas prove of permanent utility.

CHAPTER TWO

METHOD AND TECHNIC

1. Automatic Internal Dissection

By means of fine, straight and curved chisels, such as were described by the author some years ago (7), it is possible to dissect out with a tolerable degree of accuracy the cerebral tissue containing the intercortical systems. Considering the large area and the irregular conformation of the surface of the human cerebrum, the task of such dissection is rather great. Much labor and time are saved and a far greater degree of accuracy and of detail are obtained by the following procedure of automatic internal dissection by means of graded internal explosions of the cerebrum (8):

(a) The cerebrum is thoroughly fixed either in formaldehyde or in a dilute solution of alcohol. Since it is desirable to separate the bundles of nerve fibers as much as possible, the tissue must be firm and elastic, not brittle. Fixation which renders the tissue brittle, such as strong alcoholic solutions or those containing more than a trace of bichromate salts, are unsuitable. After fixation, the specimen is stripped of the membranes, wrapped in several layers of gauze, then in three to five layers of bandages. It is then placed in a strong, air-tight metal container.

(b) The container, which after a number of experiments was finally used with success, is made of steel, and its walls, its cover and its mountings are strong enough to withstand several thousand pounds of gas pressure, thus leaving a wide margin of safety. Upon the cover are mounted a gauge capable of registering 2,000 pounds and a high-pressure valve with an opening 12 millimeters in diameter. A high-pressure valve with such a large aperture ordinarily requires the complete revolution of the handle before it is either completely opened or closed—an operation which takes about one second. Since the power of an explosion is in an inverse ratio to the length of the time during which it takes place, it is desirable to open the valve completely within as small a fraction of a second as practicable. For that reason, the handle of the valve is connected by means of a set of gears to a lever in such a way that a single pull on the lever opens the valve completely.

The valve is threaded in front for a coupling with a flexible tube which leads to a compressed-gas tank. Several gases were tried and abandoned as impracticable and the one finally used successfully was liquid CO_2. The pressure of this gas varies with the room temperature between 850 and 1,200 lbs.

The specimen is subjected to the pressure of the gas for one or two days in order to allow the latter ample time to dissolve in the fluid of the brain. The valve is then quickly opened, with the result of an internal explosion of the tissue. The expansion of the tissue by the gas is, however, counterbalanced by the resistance of the bandages. Instead of disintegrating, the cerebrum is found to have been beautifully dissected along "natural" lines of cleavage. At points where there is much interweaving of fibers, the tissue remains practically intact. In the gross, therefore, the brain stem, and especially the feltwork of the cortex, remain practically uninjured, while a section through the white substance of the cerebrum discovers a rich design of astonishing regularity. Even in the gross specimen the fact becomes manifest that the white substance of the cerebrum is built out of closely apposed laminae, a number of which are no thicker than the paper of this monograph.

(c) I have immersed a number of such exploded preparations in India ink; these were then dehydrated, embedded in celloidin and cut in serial sections in the ordinary way. Sagittal, horizontal and coronal sections from different parts of the cerebrum are shown in Figures 1, 2 and 3. It will be seen that a reconstruction of a series of such sections must give the *gross form* of a number of the cerebral pathways. The words "gross form" are emphasized on account of the frequency with which a gross form of a cerebral pathway is confused with the direction of its constituent fibers, even in studies based on degeneration specimens and serial sections stained by the method of Weigert-Pal.

(d) It is generally desirable to repeat the process of exploding the specimen a number of times, in order to make sure that the dissection has been carried as far as possible. After each explosion the tissue is to a certain extent dehydrated. Water is forced out of the blood vessels and, in all probability, a small quantity from the spaces formed in the tissue by the explosion. It is very doubtful whether cells are ruptured.

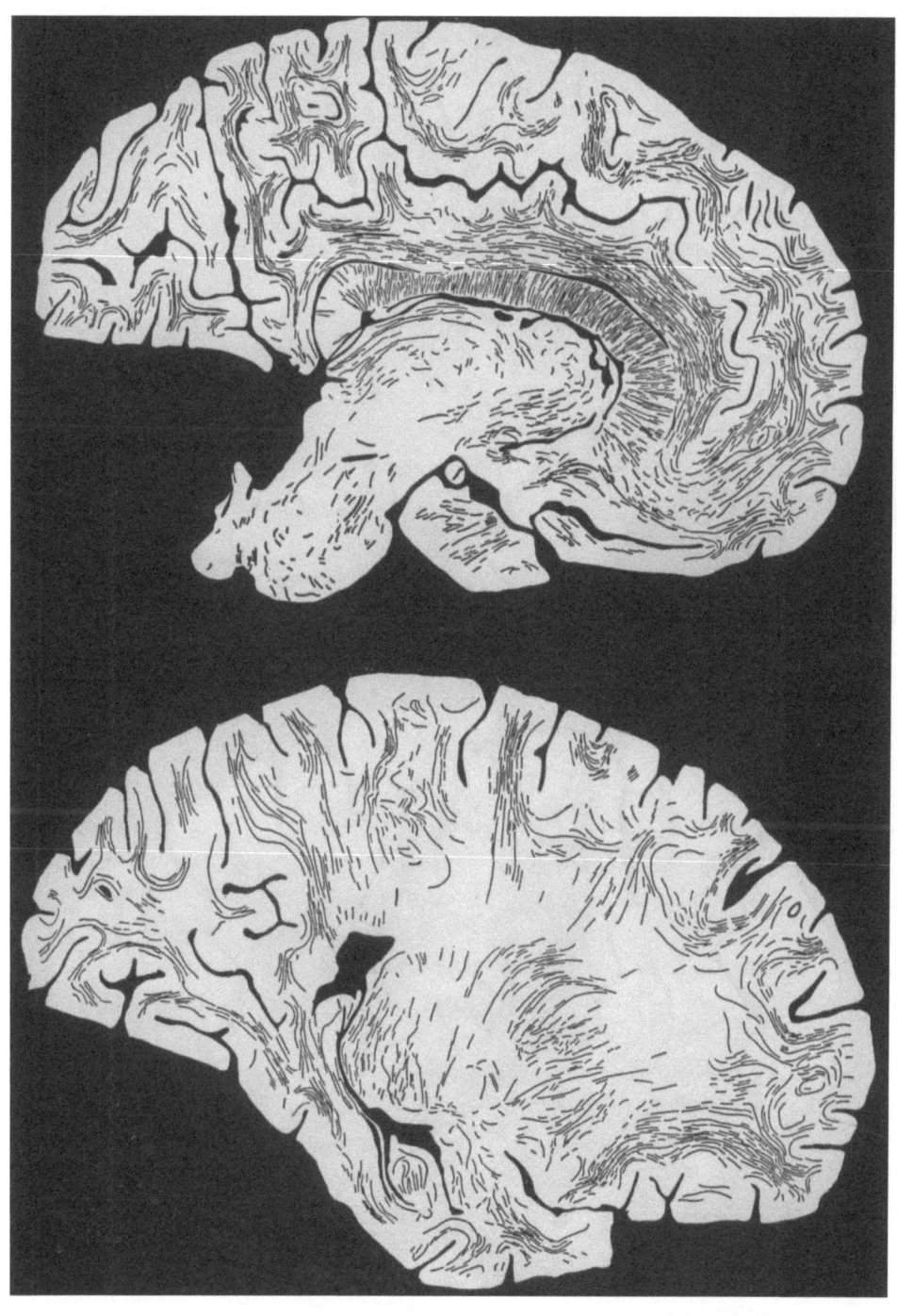

FIG. I A. SAGITTAL SECTIONS FROM DIFFERENT PORTIONS OF AN EXPLODED HUMAN CEREBRUM TREATED WITH INDIA INK

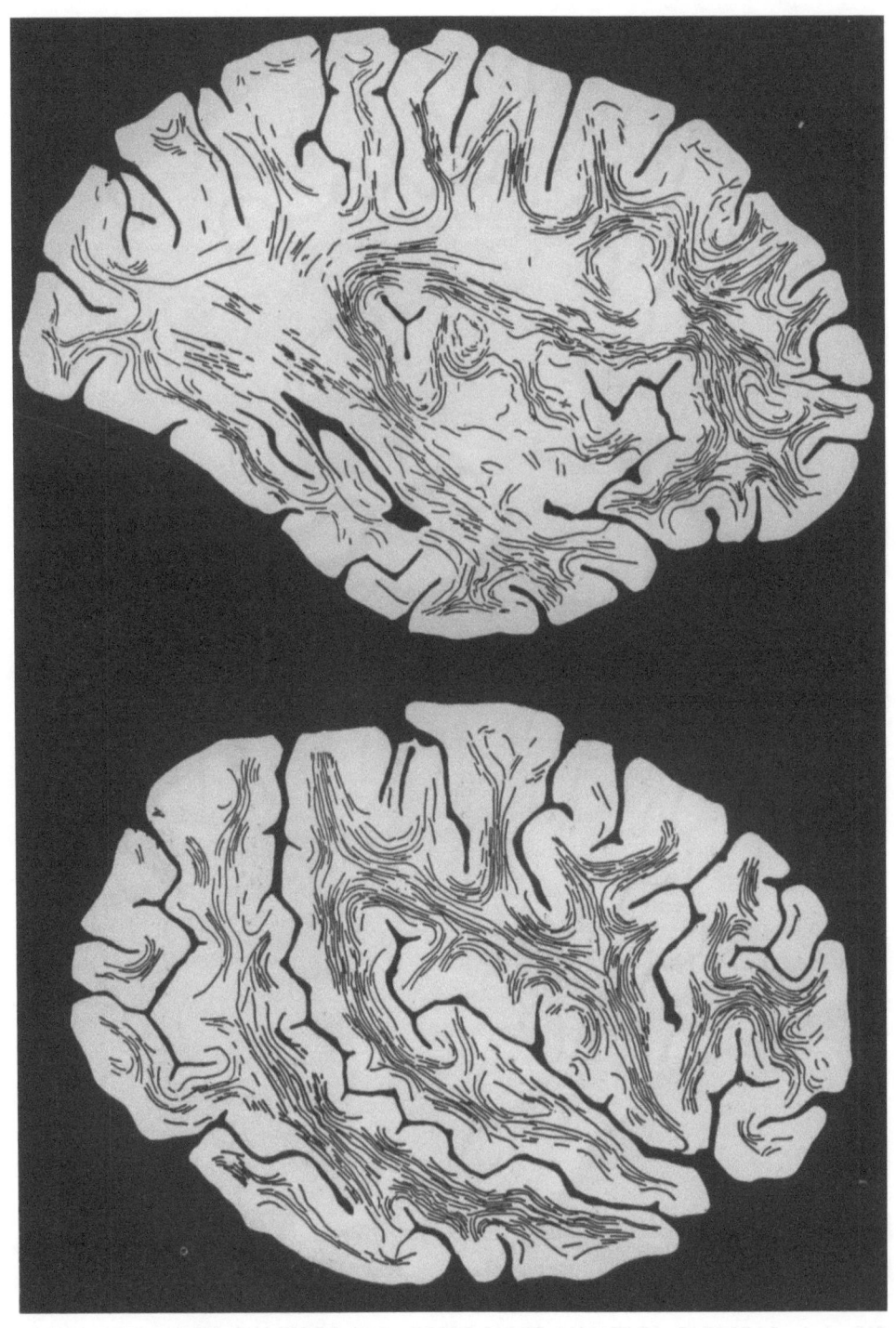

FIG. 1B. SAGITTAL SECTIONS FROM DIFFERENT PORTIONS OF AN EXPLODED HUMAN CEREBRUM TREATED WITH INDIA INK

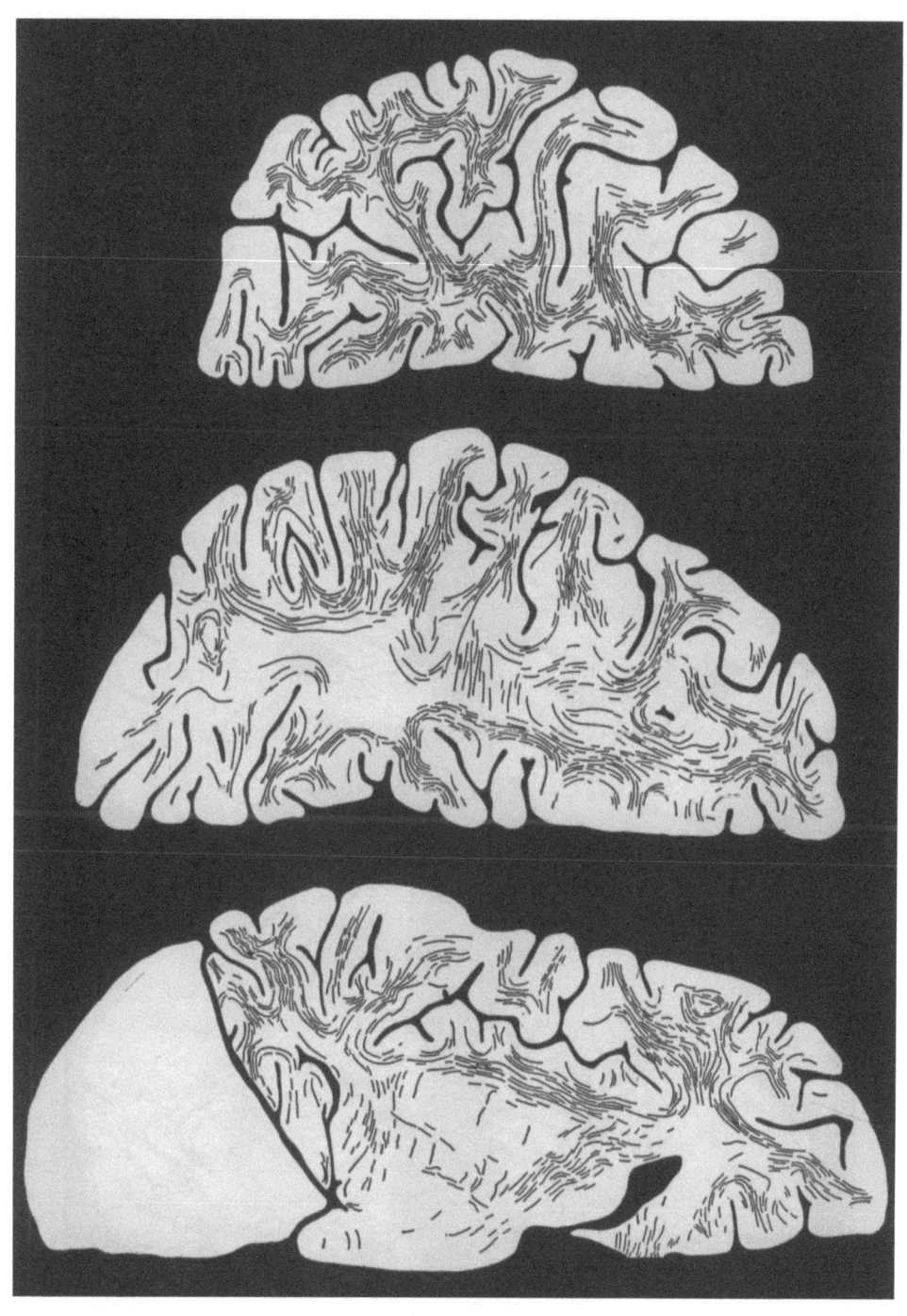

FIG. 2. HORIZONTAL SECTIONS FROM DIFFERENT PORTIONS OF AN EXPLODED HUMAN CEREBRUM TREATED WITH INDIA INK

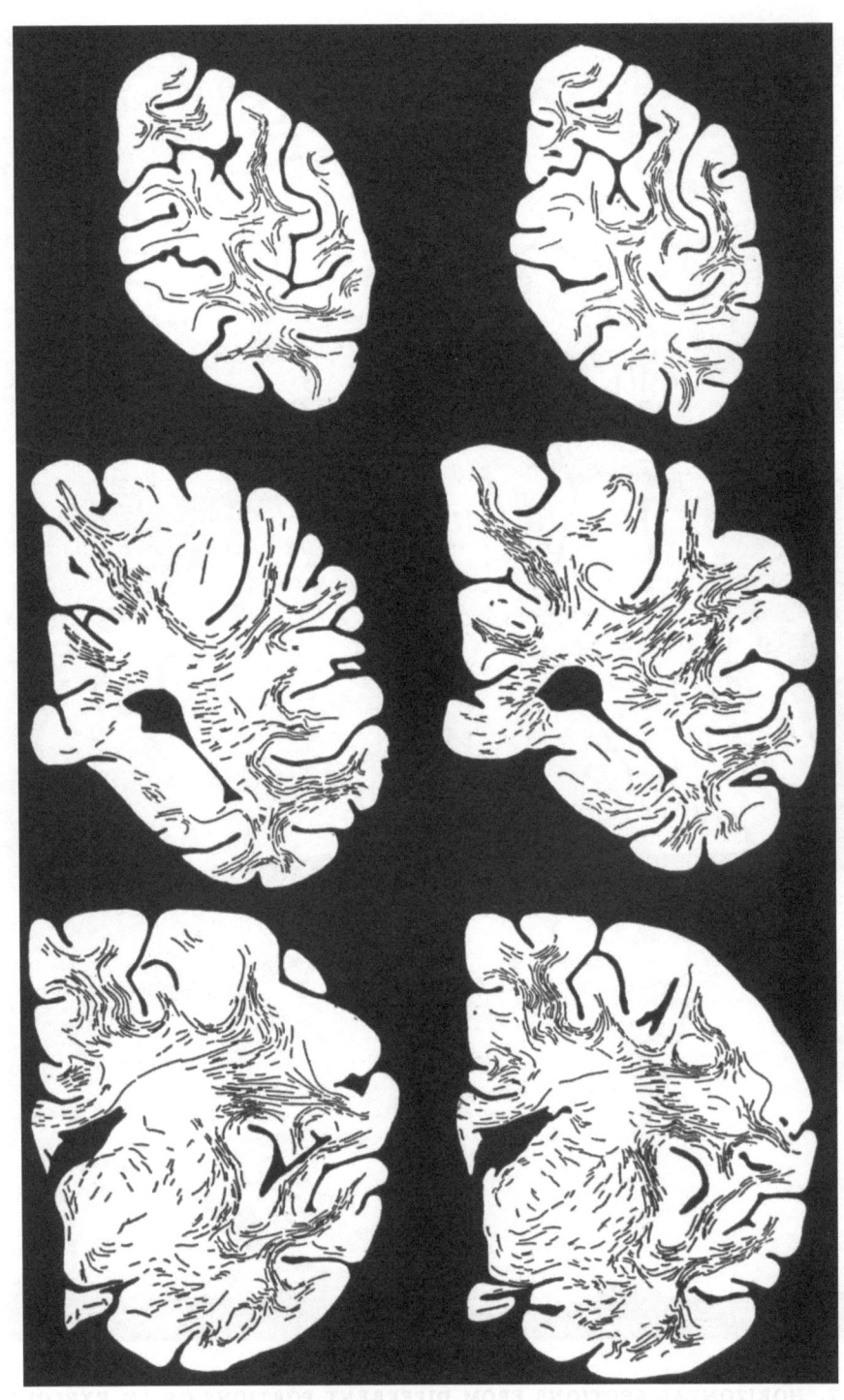

FIG. 3. CORONAL SECTIONS FROM DIFFERENT PORTIONS OF AN EX-
PLODED HUMAN CEREBRUM TREATED WITH INDIA INK

An examination of the cells by means of the Nissl and other cell stains showed a dislocation of the nucleus and other abnormalities, but nowhere a rupture. That water is not forced out except from the blood vessels and the newly formed spaces is also attested by the following fact: After the first explosion, about 10 percent of water by weight is expelled; after a second explosion, about 2 percent; and after the subsequent explosions a progressively diminishing fraction of one percent.

Since the spaces formed are mechanically dehydrated, and since the force of the expanding gas is most effective when acting against the resistance of the water in which it is dissolved, it is desirable to supply the liquid after each explosion. Immediately after each explosion I have therefore immersed the specimen, bandages and all, in boiling water, which was then allowed to cool. Three desiderata were accomplished. The residual gas was forcibly expelled while the water was hot; water entered the spaces in the subsequent process of cooling; and the tissue was further fixed and toughened, rendering it more suitable for the subsequent processes of both manual dissection and staining by the method of Weigert-Pal. In preparing the inked specimens, boiling ink was used instead of water.

2. Manual Dissection, Plaster-of-Paris and Metal Casts

A cerebrum prepared in the manner described can be dissected by hand with great ease. The white substance containing the intercortical systems underlying the fissures comes away whole at the least touch of a tool—can indeed be peeled out by the bare fingers. Actually this was accomplished by removing first the central tissue of the cerebrum and leaving the subcortical white laminae of the fissures attached to each other by the cortex and by the line of extraneous fibers—the ends of projection fibers. Further information on this point will be given subsequently in connection with the technic of obtaining preparations of the deep intercortical systems. The appearance of such a dissection is that of an obverse of the configuration of the cortex, each fissure of the cortex being represented by a protuberance, each crest of a convolution, by a recess (Figures 4, 5, 17, 18, 38).

It was important to preserve *permanent records* of these dissections before proceeding with the preparation of the blocks. Accordingly, plaster-of-Paris or metal casts were made of them; the former, from molds made by dipping the preparation in a mixture of beeswax

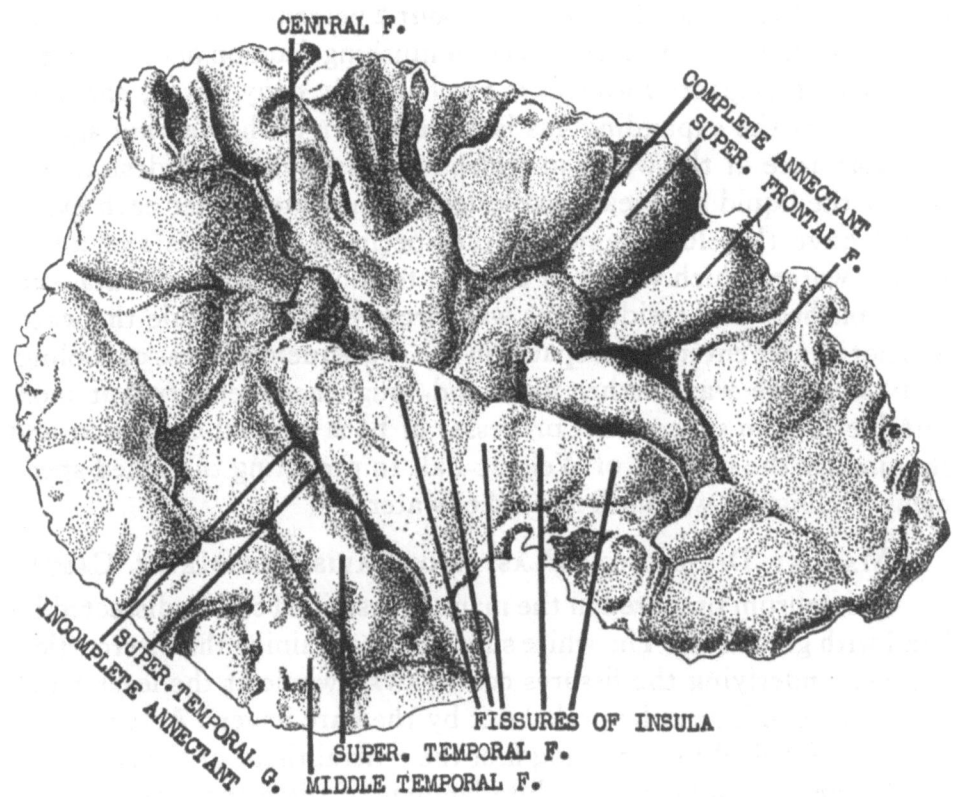

FIG. 4. THE WHITE LAMINAE UNDERLYING THE CORTEX OF THE FISSURES OF THE LATERAL SURFACE OF THE LEFT HEMISPHERE OF A HUMAN CEREBRUM

The occipital lobe had been separated from the rest. The picture presents an obverse of the configuration of the cortex. The crests of convolutions, superficial as well as deep, appear on the picture as clefts of greater or lesser depth; the protuberances on the picture are the fissures; the indentations on the protuberances are partial annectants, buttresses, etc. Two or more fissures are frequently embraced in a single boat or basketlike lamina. The drawing was made from a plaster-of-Paris cast of a dissection of an exploded cerebrum.

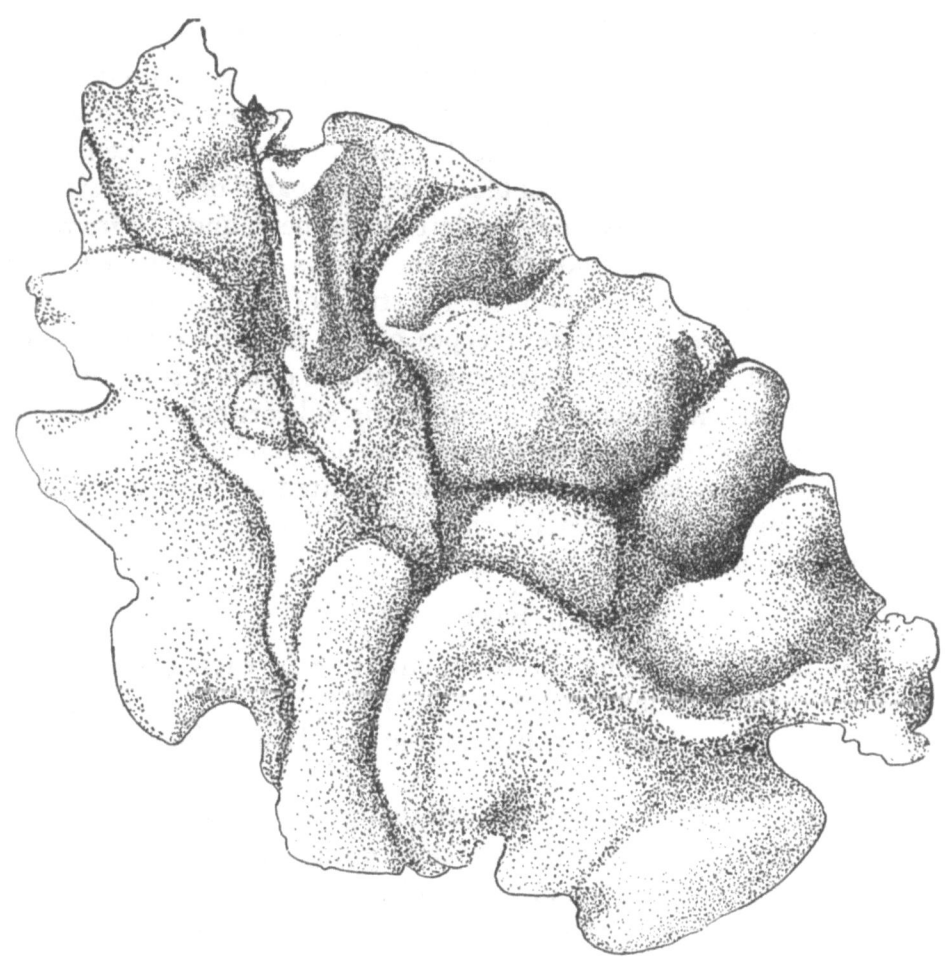

FIG. 5. THE NEXT STEP IN THE DISSECTION OF THE PREPARATION SHOWN IN FIGURE 4

The insula, the temporal and the prefrontal convolutions have been broken away from the rest of the preparation, exposing the intercortical systems of the superior operculum. The drawing was made from a plaster-of-Paris cast.

and paraffin, the latter from molds made of a mixture of plaster-of-Paris and powdered pumice stone. The most suitable metal for the purpose is type metal, which can be subsequently "flushed" with silver and laquered. *The drawings of the dissections which appear in connection with this report were made not directly from the dissections but from the casts of these.* The advantages of such a process are obvious.

The fibrous structures represented by the protuberances in such a dissection (the obverse of the cortical fissures) were then broken away from each other. Altogether 144 blocks were prepared of the intercortical laminae of the fissures from the left hemisphere, and 113 from the right, of one brain. A number of other blocks were prepared from different portions of other brains for the study of the intercortical systems of the fissures in different types of the human cerebrum. The deeper intercortical systems—the entire capsule of the insula, the vertical fiber sheet which extends lateral to the subcortical fiber systems of the parieto-occipital and the calcarine fissures, and others, were prepared in large blocks and the sections mounted on correspondingly large glass slides. The special technic by which these fiber systems were obtained will be described later.

3. Recording

The task of recording the exact situation of each of the many blocks of the subcortical laminae of the fissures on the surface of the cerebrum is by no means a contemptible one. It was done in the following way:

A plaster-of-Paris cast and line drawings were made of each hemisphere (Figures 6, 19, 21, 33). As each piece of tissue was taken out of the general subcortical shell of white substance (Figures 4, 5), an outline of that piece in its proper place was marked with India ink on the plaster cast and on the line drawing, and the circumscribed space was marked by a number which corresponded to the number of the piece of tissue. Holes were then punched in the piece of tissue, near the edges, and incisions were made along certain lines for reasons to be explained in the next paragraphs. In order to prevent the turning over of a section in the subsequent processes, the thin block of tissue was marked with three notches, one on the right-hand edge (facing the hemisphere in a given way) and two on the lower edge. The holes, cuts and notches were marked in their respective places on the plaster-of-

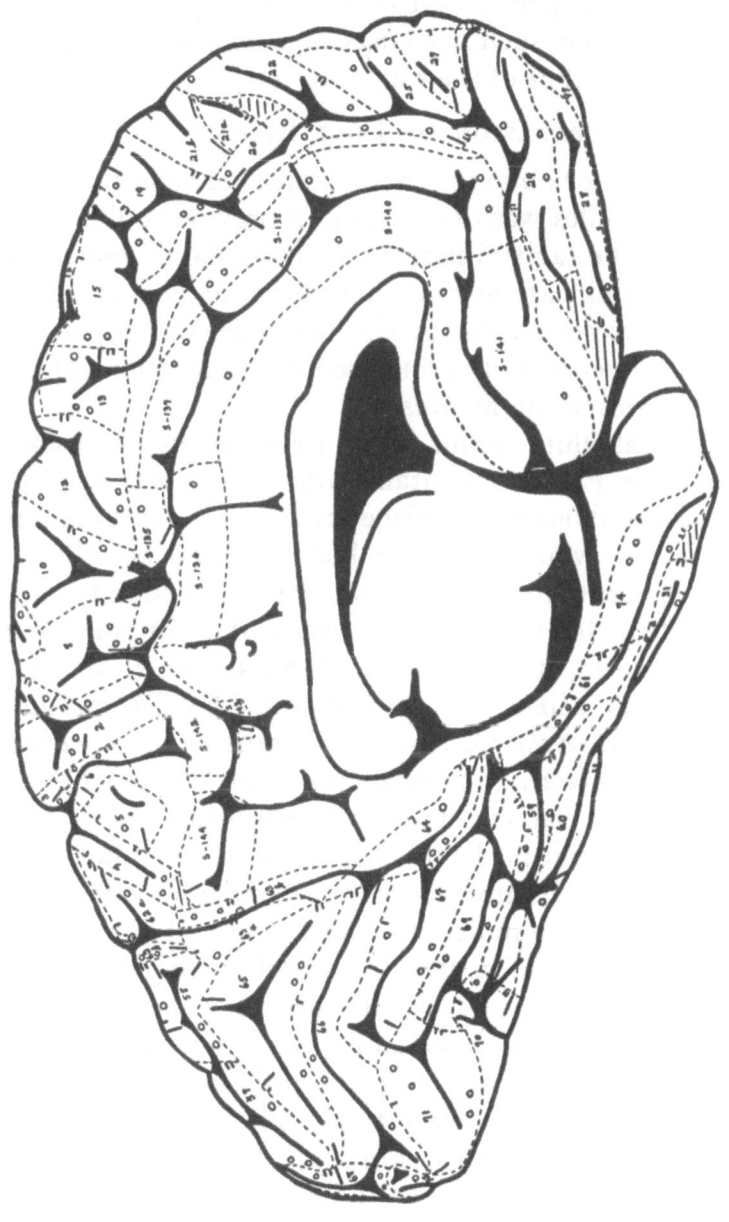

FIG. 6. A LINE DRAWING OF THE MESIAL SURFACE OF THE PLASTER-OF-PARIS CAST OF THE LEFT HEMISPHERE

The numbered spaces circumscribed by the dotted lines represent the blocks of tissue removed for further treatment. The circles represent the holes punched in the tissue in the same situations; the long single dashes perpendicular to the dotted lines are the cuts made in the boatlike structures; the short double dashes represent the notches punched on the edges of the laminae in the same situations. All these markings were made on the plaster-of-Paris casts as well as on the line drawings.

Paris cast and on the line drawings. A separate outline of the flat block with its several markings was kept in the daily journal. If, during the process of sectioning, it happened that only a part of a section was shaved off by the microtome, such markings as it bore were sufficient to locate the part on the outline of the whole section in the journal.

4. The Preparation of Flat Blocks

The general method for obtaining microscopic sections along the course of nerve fiber bundles was described in a previous publication (7) in connection with the sagittal portion of the thalamic radiation. Slight but important modifications have since been made in the technic. It will therefore be expedient to describe the process briefly in this place, with especial reference to the microscopic preparations of the intercortical fiber systems of the fissures.

(a) The subcortical white lamina, of an irregular boat or basket shape, such as shown in Figure 7, contains intercortical nerve fibers. Its average thickness is about one millimeter. In order to obtain microscopic preparations in which the fiber bundles could be followed along their course, the bent lamina which is to be sectioned, must first be flattened. Owing to its boatlike shape, however, it is not possible to flatten it without fracture; it is therefore advantageous to incise it along lines where a break is likely to occur (Figure 8 B). *The incisions in the lamina are recorded on the plaster-of-Paris casts and in the drawings in the corresponding places.* The cortical tissue on the inside of the boatlike lamina, being less elastic than the white substance on its outside, is almost certain to split in the midline when the lamina is flattened and to involve the white substance in the damage. It is therefore advisable to scrape away the more superficial layers of the cortical substance. The fiber side (that is, the outside) of the lamina is then placed on a glass slide; it is covered on the cortical side with a thick pad of gauze which is, in its turn, covered by another glass slide (Figure 8 C). The two glass slides, with the lamina of tissue and the gauze pad sandwiched in between, are then tied together with a strong string, which is wound around a number of times, more or less tightly, until the white side of the lamina is seen to be evenly applied against the glass.

Previous to flattening the lamina, it is marked with a number of holes and notches, which are recorded in the drawings and on the casts in the corresponding places.

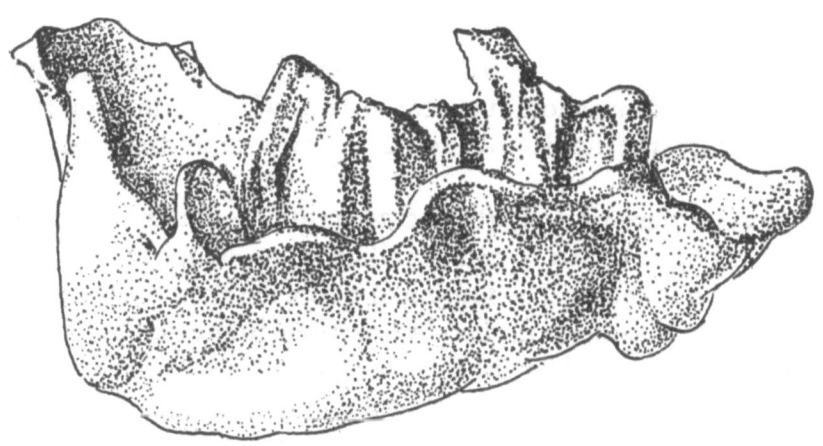

FIG. 7. A BOAT OR BASKETLIKE SUBCORTICAL WHITE LAMINA OF A FISSURE

Drawing from a metal cast of a dissection

(b) The package is then placed in the mordanting solution (for the Weigert-Pal stain) for two or three days, after which it is untied and the pad of gauze and the glass cover are removed. A single layer of gauze is placed between the flat lamina and glass slide on which it rests (Figure 8 D). The lamina, gauze, and glass slide are tied by a few windings of string—rather loosely, or the surface of the tissue will bear the imprint of the gauze which is placed between it and the glass.

The purpose of the layer of gauze between the lamina and the glass is to make the fiber surface of the tissue more accessible to the mordanting solution than must be the case if it rested flat against the glass.

The package is then replaced in the mordant.

(c) The processes of mordanting, washing, dehydrating and impregnating with thin celloidin are carried out without untying the preparation. Before the final embedding in thick celloidin, one proceeds as follows:

A piece of pure tin foil, about 0.25 millimeter in thickness, is flattened by rolling it with a round pencil on glass. Much thinner foil may not retain its flat shape. The lamina of tissue is taken off the glass slide and the gauze is removed; the glass is then immersed in the dish of thick celloidin, followed by the flat piece of tin and the lamina of tissue, which are placed on top of each other in succession. The lamina is gently pressed down on its bed with a glass rod (Figure 8 E).

The reason for immersing the glass, the tin and the tissue one after another in succession rather than in one package is in order to exclude the air between them, which, when present, is almost certain to lift the lamina from its bed at one or more points and so distort its even plane.

(d) After cutting the preparations out of the thick celloidin (a sharp tool bent at a right angle passed under the glass slide will lift it off the bottom of the dish without difficulty), it is mounted on a fiber block in the usual way, but without removing either the glass slide or the tin. When mounted on the fiber base, the relations are in the following order (Figure 8 F): fiber base, a layer of celloidin, the embedded flat lamina of tissue, heavy tin foil, glass slide. The entire block is then placed for twenty-four hours or more in 90 percent alcohol.

Under no circumstances should the glass slide and the tin foil be

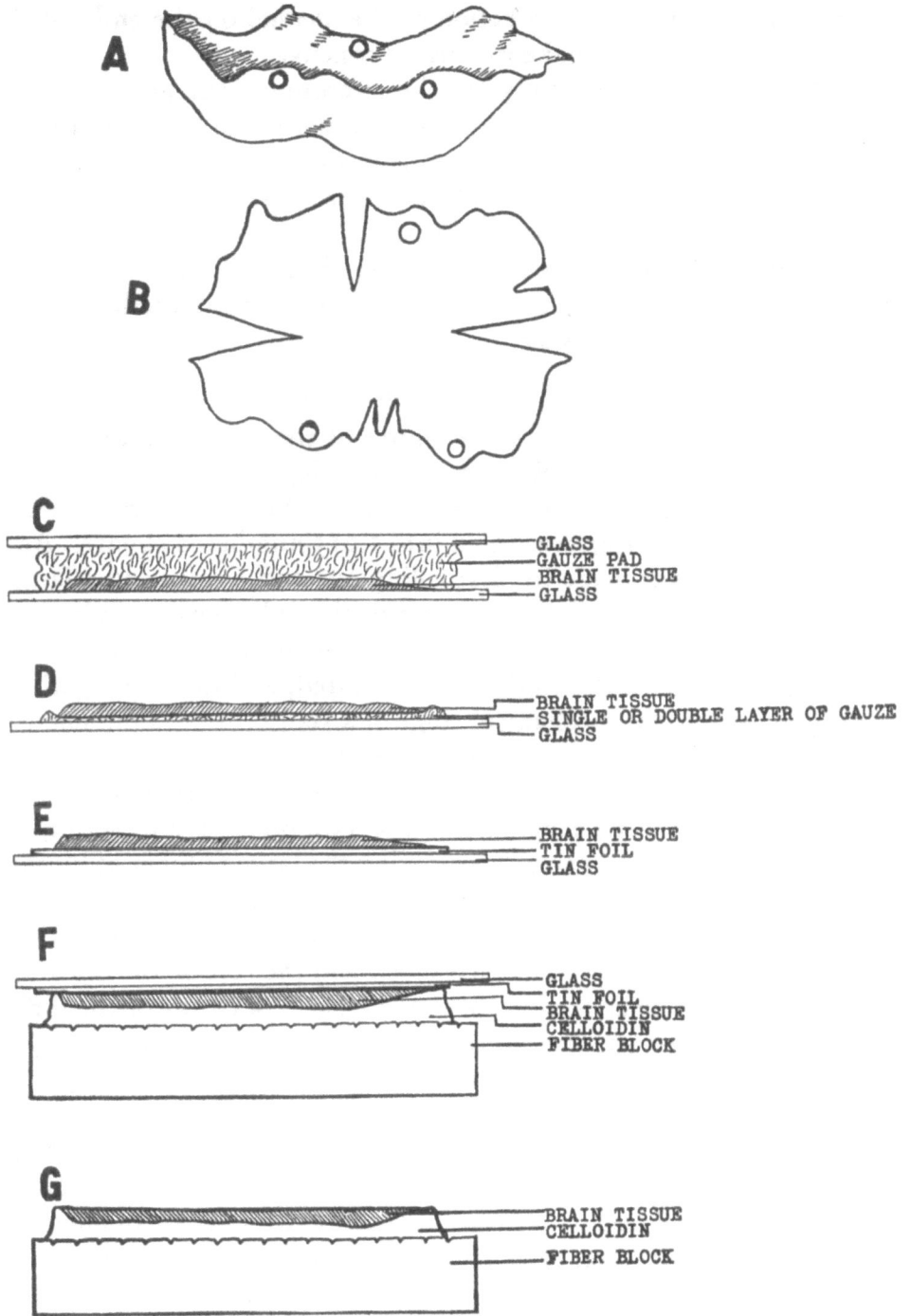

FIG. 8. A DIAGRAMMATIC REPRESENTATION OF THE SEVERAL STEPS IN THE PREPARATION OF FLAT BLOCKS, EXPLAINED IN THE TEXT

removed at this stage, or the action of the alcohol on the embedded lamina will be certain to make its surface concave.

(e) When the block is ready for the microtome, the glass slide is removed by inserting the edge of a safety razor blade between it and the tin foil. An edge of the tin is raised and rolled on a round pencil off the lamina of tissue. The surface of the latter is found to be a straight plane, glistening like a mirror. It (Figure 8 G) is then adjusted in the microtome, as far as possible in such a way that the first cut will produce a complete section. The sections are then remordanted, stained and mounted in the usual way.

5. Detailed Drawings

The drawings which show the course of the fiber bundles of the intercortical systems, were made in the following way:

A glass slide of the same size as those on which the sections were mounted (2 by 3 inches), was ruled into 1,015 numbered squares. Actually this slide was made by first ruling out and numbering a large sheet of paper with India ink and then reducing it by means of photography on glass. The ruled slide was fastened by means of adhesive tape to that on which the section was mounted, and the two together were inserted into the mechanical stage of a Bausch and Lomb microprojector. The drawings were made at magnifications of 50 and 150. It will be readily seen, however, that a section measuring, say 5 x 5 centimeters, drawn at a magnification of 150, must give a picture of about 7.5 x 7.5 meters in size. Considering the large number of blocks and the fact that from 10 to 40 sections were cut from each block, such a task was impractical, even though the largest part of a single section might actually be drawn from each block and the rest of the drawing filled in from the other sections of that block. A way out of the difficulty, with as little sacrifice of accuracy as was consistent with the nature of the enterprise—to trace the apparent course of the pathways—was found in the following method: Sheets of paper, ruled into 1,015 numbered squares, each square 8 x 8 millimeters, were printed. The contents of each of the microscopic squares on the ruled glass slide, as they appeared at a magnification of 50 and 150 diameters, were drawn free-hand on each of the 8 x 8 millimeter squares on the ruled paper. The possible error must be very slight, since the fiber bundles extend from any one square into the adjoining squares. The

FIG. 9. A PHOTOMICROGRAPH (× 75) SHOWING TWO SETS OF INTERCORTICAL FIBER BUNDLES CROSSING EACH OTHER AT RIGHT ANGLES, IN THE MANNER OF A TEXTILE FABRIC, IN THE SUBCORTICAL LAMINA OF A FISSURE

lines drawn in each square served, therefore, to a certain extent as a control of the direction of the lines drawn in the adjoining squares. In this manner the microscopic image of the section was at once reduced within practical dimensions.

One of the main difficulties lay in the fact that patches of cortical substance interrupted the picture of the pathways in a number of sections, frequently making it impossible to produce a complete drawing from a single section. Most of the detailed drawings, therefore, such as shown in Figures 17, 24, 27, etc., represent reconstructions from a number of serial sections.

After drawings were thus made from sections of all the blocks, there remained the problem of joining them in a natural way so as to represent the course of the nerve pathways as they are laid down in the cerebrum.

6. The Preparation of Schemata from Detailed Drawings

The work of fitting the detailed drawings of the intercortical systems of the fissures into a schematic picture of the hemisphere, requires a considerable amount of experience. When a single curved lamina is cut in two or three pieces and each piece flattened separately and cut into sections, the microscopic drawings of the different parts of the lamina do not readily connect into a single congruous design. If a design be drawn on the peel of an orange and the peel then be cut in pieces and each piece flattened, the task of composing the former entire design out of its distorted parts may be fraught with serious difficulties. In the present work it was accomplished as follows:

The schematic outline of the partly opened fissure was marked with the numbers, the circles, notches and wedges in exactly the same situations as they appeared on the line drawings and the plaster-of-Paris casts. It will be remembered that these different markings represented the holes, notches and cuts on the blocks of tissue made for purposes of orientation. In transferring the lines from the detailed separate drawings on to the schematic picture of an entire fissure, *the worker adhered rigidly to the relationship between the lines and the various landmarks, disregarding completely the relation of the lines to the points of the compass.* After a few days' work one acquires the habit of disregarding the apparent directions of the lines and of adhering rigidly to their relations to the landmarks. The work then proceeds with dispatch.

This final reconstruction is one of the most important and, for the beginner, one of the most difficult items of the technic. It will therefore repay the worker who might wish to pursue this line of research to study the reconstruction of the central fissure from the detailed drawings of the three blocks into which its subcortical lamina was cut, given in Figures 24, 25 and 26.

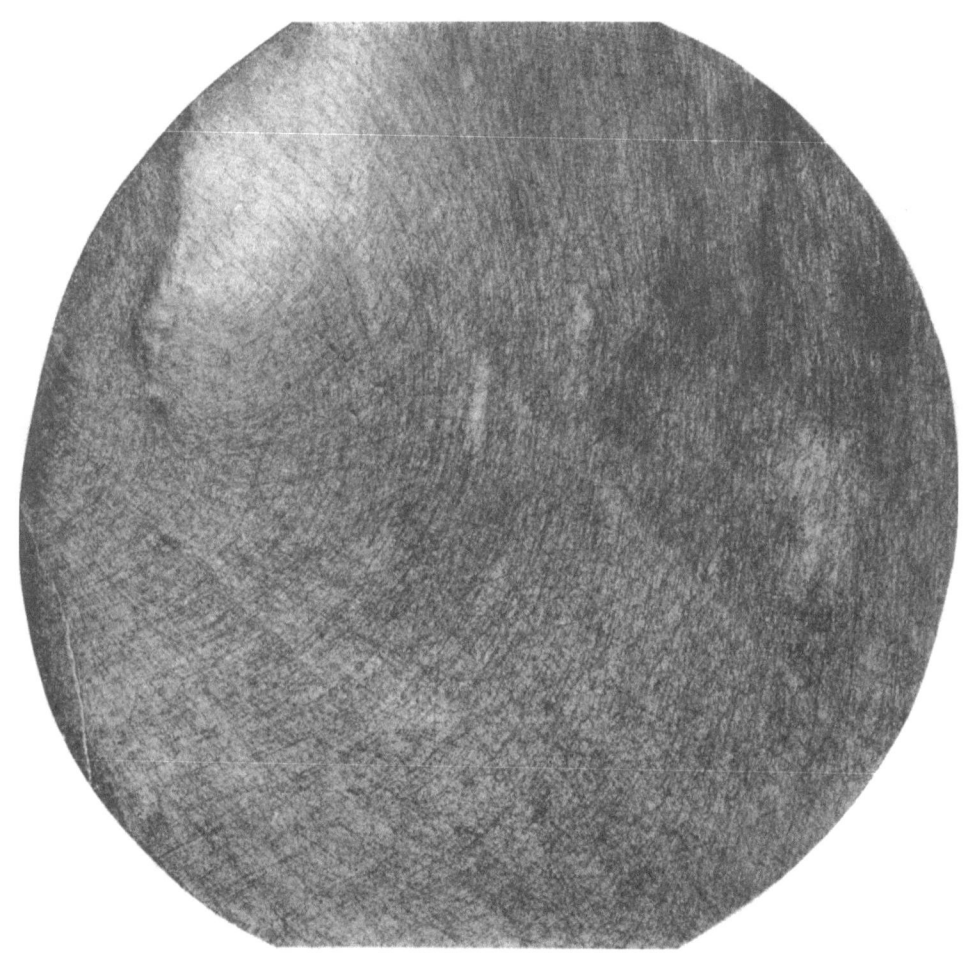

FIG. 10. A PHOTOMICROGRAPH (X 35) SHOWING CURVED INTERCORTICAL PATHWAYS FROM A SUBCORTICAL LAMINA OF A FISSURE

CHAPTER THREE
THE SCOPE OF THE WORK

At a magnification of about 300 diameters, the microscopic field being small, but few of the fibers in the field are observed to pursue parallel courses. At a magnification of from 75 to 150 diameters, in the sections prepared by the methods described, the microscopic field is seen to be covered by one, two or more sets of straight or curved parallel fiber bundles which cross each other at different angles (Figures 9, 10, 11). If one can imagine a kind of textile fabric which consists not only of two sets of parallel bundles of threads which cross each other at right angles, but of several such sets crossing each other at different angles, he will have a picture which is nearly like that seen in these sections.

A closer study reveals two disconcerting facts. One is that in and between the parallel fiber bundles there are large numbers of single fibers which do not pursue a course parallel to any of the fiber bundles in the field. The latter constitute an irregular network of single fibers which is spread out tangentially over the entire extent of the cortex. It is strikingly like that of the intercortical network of the opossum, shown in Figure 39. Visualized as a whole, the tissue from which these sections were cut might be compared to a dense but orderly textile fabric permeated by a rather sparse feltwork of single fibers of inconceivabe complexity. Another disconcerting fact is the one to which attention was called in a previous publication (7): Although the microscopic field at first sight appears to be covered by a sheet of parallel fiber bundles, a closer examination reveals fibers issuing at short intervals from any one bundle and entering the adjoining bundle (Figure 11), so that it becomes impossible by the method employed to trace individual fibers for any great length. An examination of the India-ink sections made from exploded brains, shows throughout frequent interruptions in the black lines. The intervals between those lines represent the tissue held together strongly enough to resist the force of the internal explosion—they are points at which much interweaving of fibers takes place. That a number of fibers leave the bundles between the latter points and that they are torn by the explosion is certain (Figure 11), but that number is at any rate small as compared to the other.

It goes without saying that the method employed is not calculated to give any information regarding the direction of the nerve impulse, that is, regarding the situation of the cells of the respective axones.

The present work is therefore limited in its scope to the mapping out of the general direction of the majority of parallel fiber bundles which appear to proceed from certain points in the cerebral cortex to others, in microscopic sections prepared from flattened dissections.

FIG. 11. A PHOTOMICROGRAPH SHOWING THE ARRANGEMENT OF THE FIBERS IN THE BUNDLES AND THE DEFECTS PRODUCED BY THE EXPLOSION OF THE CEREBRUM

At several points fibers may be seen to leave some bundles and to join others. Fiber bundles broken by the explosion may be seen in the upper part of the photomicrograph $\times$ 35.

CHAPTER FOUR

REGARDING THE CELL ORIGIN OF THE INTERCORTICAL SYSTEMS

A century of observation and investigation of these structures has brought no definite information regarding any single type of cells in the cortex from which they may originate. Meynert (3) ascribed the source of the U-fibers to the fusiform cells of the lowest cortical stratum. His conclusion is in the nature of a theoretical deduction from the relation of the long axis of the fusiform cells to the curvature of the cortical line of the fissures. The hypothesis is rather too distant to be accepted for a fact. Sections carried squarely across a fissure and stained by any of the current methods show the curved fibers immediately underlying the cortex of the fissures to penetrate it to varying depths. A certain number of them, to be sure, are lost to view in the lowest cortical stratum, but most of them can be traced to more superficial strata, among the large and medium-sized pyramidal cells.

Cajal maintained emphatically that most, if not all, the intercortical fibers are collaterals of projection fibers which are derived from pyramidal cells of all sizes.

I believe [says he (9)], as a result of my latest investigations, . . . that most of the projection fibers send out association fibers for long distances and that a good many of the association systems described by authors pass not directly and exclusively between two areas, but that they are bundles of collaterals or branches which arise from projection fibers whose main branch, incorporated in the corpus striatum, constitutes a motor pathway.

And again (10):

The latest investigations of the cortex of the rabbit, rat and especially the seven to twenty day mouse, have convinced me of the important fact that most or even all of the homolateral association fibers which pass out of the sensory areas of rodents, do not constitute direct pathways, but that they are collaterals or branches of projection axis-cylinders. At least this is the ordinary fact regarding the association fibers in the motor, optic, interhemispheral and sphenoidal olfactory regions, etc.

There is no reason to doubt that some of the intercortical fibers in the *human cerebrum* arise in that way; but the universal origin of the intercortical systems as collaterals from projection fibers, is, to my mind, quite unlikely, for these reasons:

As far as is known, the projection systems of the cortex consist of the cortico-pontine, cortico-spinal and cortico-thalamic fibers. The former two are aggregated in the pes pedunculi, whose area in cross section hardly exceeds 10 square millimeters. The thalamic fibers have their greatest aggregation about the lateral wall of the thalamus. The thickness of the thalamic fan in that situation hardly exceeds 2 millimeters and the length of the curved line of emergence or entry of the fibers is about 75 millimeters. The area of cross section of all the fibers which enter and leave the thalamus is, therefore, roughly about 150 square millimeters. Allowance is made for a very large margin of error by granting that the cortico-thalamic fibers, in cross section, take up half this area. That being granted, the sum of the cross sections of all the projection systems of a hemisphere of the human cerebrum is about 85 square millimeters. I have measured the plane of junction of the cerebral cortex with the underlying white substance by multiplying the perimeter of a set of serial sections of the cerebrum by their thickness and found it to be about 60,000 square millimeters. Only half of this fibrous extent can be considered as the area of the sum of the cross sections of cortico-fugal fibers of all kinds; and of this, only 1/353 part (30,000:85 = 353) is occupied by the projection systems. If all the intercortical fibers are collaterals of projection fibers and if the thickness of the collaterals is the same as that of the parent axones, then each projection fiber must give off 353 collaterals immediately beneath the cortex. This is not the case.

In justice to Cajal's observations, however, it must be borne in mind that the adult human cerebrum, with which we are dealing, is a vastly different structure from that of the young rodent. With every ascent in the scale of mammalia the proportion of intercortical to projection systems in the cerebrum becomes larger. The intercortical systems are generally disposed parallel to the cortex; and so much smaller is the surface area of the smooth cerebrum of the young rodent than that of the convoluted human cerebrum, that the numbers of intercortical fibers in the small, smooth cerebrum may not exceed very much those of the projection fibers. With respect to the small, smooth-brained mammals, therefore, Cajal's observation becomes quite plausible. It is indeed possible that as in the course of cerebral evolution the intercortical systems increase in number and in bulk, the processes of specialization and differentiation are manifested in the appearance of cells whose axones are entirely devoted to the special function of intercortical connections, where before a single cell sufficed for both a projection and an intercortical fiber.

CHAPTER FIVE
ANNECTANT CONVOLUTIONS

In order to understand the course of the intercortical pathways of the fissures as they have been mapped out in the present study, attention must be called to their relation to those protuberances of the cortex within the fissures, which are known as annectant convolutions. From the point of view of the gross topography of the cortex, the annectants serve to bridge the fissures; from that of the intercortical systems, they serve as partial interruptions of the course of these systems. From the latter point of view the annectants may be divided into three classes.

The first consists of convolutions which are in no way different from those exposed on the surface, except that they are more or less concealed within the fissures. Heschl's (11) protest against calling the annectants anything but deep convolutions is entirely justifiable in the case of these structures. The relation of the intercortical fibers of the fissures to such deep convolutions is the same as it is to those exposed on the surface of the cerebrum.

Another class of annectants consists of protuberances which rise from the floor of the fissure to a variable height—about one to 6 millimeters. They seldom cross the fissure from wall to wall, generally extending from the deeper part of a wall for a variable distance across the floor, thus being separated by a groove from the opposite wall. Whether larger numbers of projection fibers enter in the composition of such protuberances of the cortex than enter the other parts of the fissure, I am not prepared to say. That such protuberances serve largely as interruptions in the course of the intercortical fibers, is certain. The long boat or basketlike dissections which contain the intercortical systems of the fissures are very apt to separate or break along lines marked by these protuberances into smaller boat or basketlike forms. However, although the majority of the fibers in question are interrupted in these protuberances, others continue their course, both by the side of such annectants and beneath their bases.

The third class of annectant convolutions consists of those protuberances from the walls of the fissures which are sometimes called buttresses. Each fits into a corresponding depression in the opposite wall of the fissure, forming interdigitations. The extent to which they interrupt the course of the intercortical pathways is in proportion to the degree to which they elevate the floor of the fissure.

Since a large number of the fibers of which the intercortical pathways of the fissures are constituted, pass beneath the bases of the smaller annectants and buttresses, one can readily see the reason why the fibrous surface of the boatlike structures obtained by dissection— the hulls of the boats—do not entirely correspond to the surface of the cortex, as may be judged by a comparison of Figure 4 and Figure 21.

CHAPTER SIX

INTERCORTICAL SYSTEMS OF THE MESIAL SURFACE OF THE CEREBRUM

A NOTE REGARDING THE METHOD AND NOMENCLATURE EMPLOYED IN THE ENUMERATION OF THE PATHWAYS

In the description of the general course of the intercortical pathways of the fissures, their massive interruption by the annectant convolutions has been to a large extent ignored and the interested student must consult the maps (Figures 12, 15, 20, 22, etc.) for such detail as is not given in the text.

In giving the course of a pathway, the expression "from X to Y" might be understood as implying the origin and termination of the nerve fibers. Since these are not known, the expressions "*between* X and Y," or "between X and (a), Y; (b), Z, etc." have been employed throughout.

The small sulci and offshoots communicating with the larger fissures play a great part in the complication of the pattern of the intercortical pathways. In order to make the complicated patterns understandable, at least decipherable on the maps, it was necessary to preface the enumeration of the pathways with a brief description of the general and special topography, for purposes of orientation. A few points regarding the morphology of the larger fissures are given in order to afford a broader view of the subject dealt with. For greater detail the reader is referred to the extensive morphological literature.

To further facilitate the reading of the maps, the groups of pathways of different directions in each fissure are numbered in the text. As a matter of convenience several such groups have been sometimes arranged under a single number.

1. THE PARACINGULATE FISSURE (FIGURES 6 AND 12)

In the hemisphere pictured this fissure consisted of a complicated plexus of sulci. It is an example of the difficulty, perhaps the inconsistency, of considering the intercortical pathways from the point of view of any one fissure. The pattern of the pathways underlying the

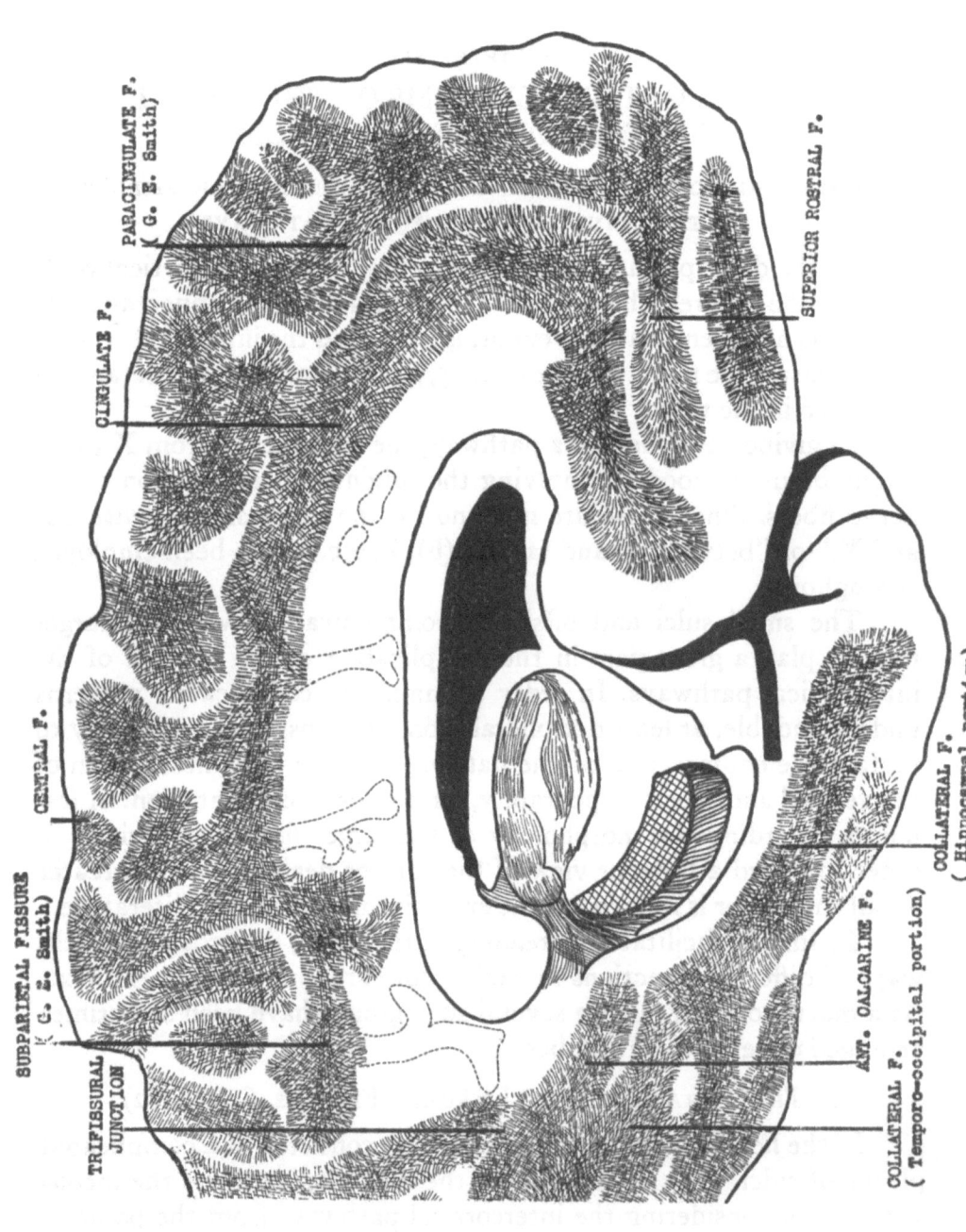

FIG. 12. A COMPOSITE SCHEMA OF THE INTERCORTICAL SYSTEMS OF THE LEFT MESIAL AREA OF A HUMAN CEREBRUM

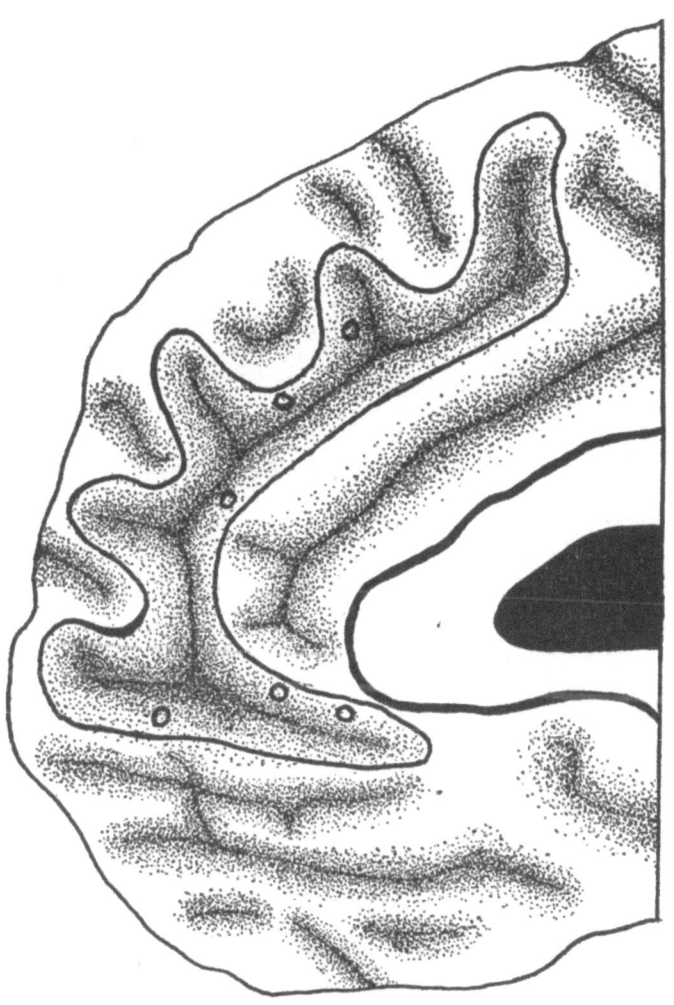

FIG. 13. A SIMPLER TYPE OF THE PARACINGULATE FISSURE THAN THAT SHOWN IN FIGURE 12

Drawing from a plaster-of-Paris cast

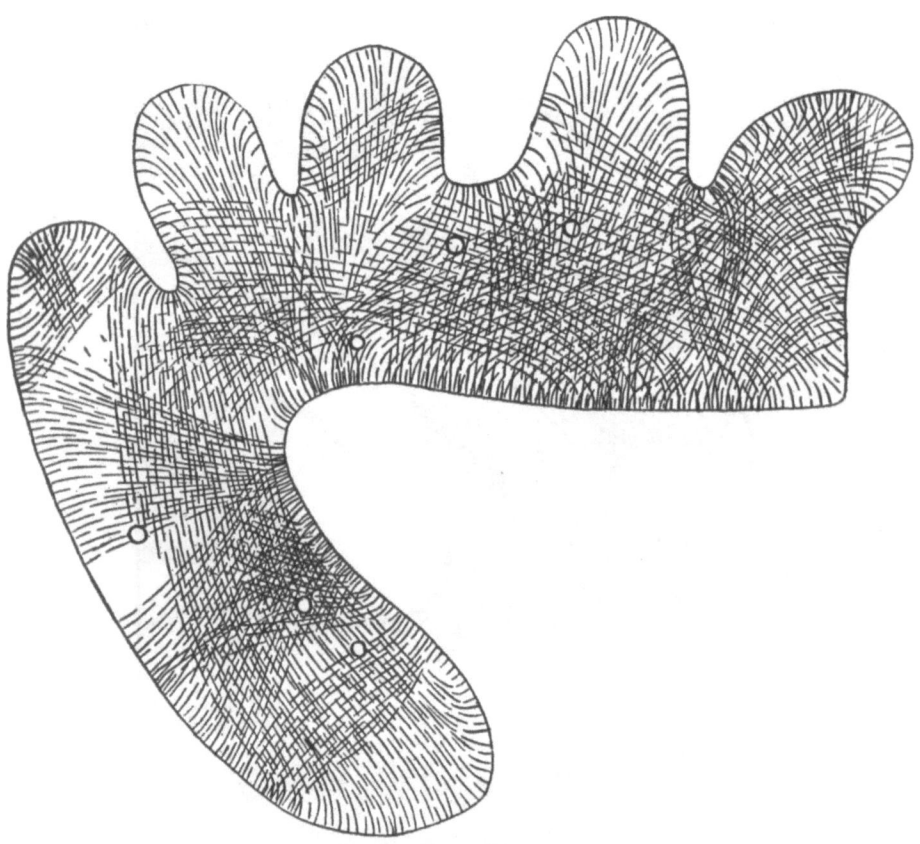

FIG. 14. THE INTERCORTICAL SYSTEMS OF THE PARACINGULATE FISSURE SHOWN IN FIGURE 13

A rational point of view (see the last chapter of the text) is the following: the pattern of the intercortical pathways of this fissure is relatively simple and the type of the fissure is in accordance with this fact.

cortex of this paracingulate fissure is apparently complicated by the superposition of pathways arriving from, or destined for, small offshoots, tributary and communicating sulci. It must be obvious that such a consideration deprives the complicated pattern of these nerve structures of much of its significance. That this point of view will in time be changed so that these pathways can be described as establishing connections between definite cell groups and areas of the cortex, is certain. The paucity of data, however, compels for the present a grouping of these pathways under the readily available, though unsatisfactory, headings of certain fissures.

Another type of the paracingulate fissure is shown in Figures 13 and 14. The number of offshoots of this fissure is smaller and the pattern of the contained pathways is therefore correspondingly simpler. From the contents of the last chapter of this report it will be seen that a more rational consideration of the relations of the intercortical pathways to the fissures calls for a transposition of the parts of the preceding sentence in this paragraph.

2. The Cingulate Fissure (Figure 12)

In view of the great age of the cingulate fissure, which had its origin in that early period of evolution when a new forebrain began to be superimposed upon the archaic rhinencephalon, one might surmise that it would appear correspondingly early in the human fetus. This is not the case. Transitory furrows and indentations, it is true, foretell the site of the future fissure as early as the fourth month of fetal life. They disappear, however, in the fifth month, leaving the mesial surface of the cerebrum smooth; and it is not before the sixth month of fetal life that several grooves and impressions appear which, by their various coalescence, build up the particular type of the permanent furrow. The reason for such a tardy development of an ancient structure is suggested by the following facts:

The rhinencephalon, or the limbic lobe of the lower mammalia, is a relatively compact structure. The unobtrusive band of the corpus callosum perforates it transversely from side to side. With the rise in the mammalian scale, there takes place an unequal increase in the size of the different parts of the cerebrum, such that the older rhinencephalon lags behind. The corpus callosum, which increases in size in proportion to the increase of the newer parts, distends and distorts the original form of the rhinencephalon. This is evident in the course of the development of the human fetus. In the young human fetus the corpus callosum and the anterior commissure appear in the form of two transverse bands of nearly equal size in front of the lamina terminalis. With the continued growth of the prefrontal, frontal, parietal, temporal and occipital lobes, the corpus callosum becomes many times larger than the anterior commissure, which remains no thicker than a match stick. The lobe of the rhinencephalon is so distorted by the corpus callosum, which perforates it, that in the adult human brain it is a matter of considerable difficulty to piece together its several parts into a single organ. Yet that is not all. Not only do the newer parts of the cerebrum outstrip in bulk the older rhinencephalon, but they appear to develop in large part at its expense. The primary receptive organ of the rhinencephalon—the olfactory bulb—is not only relatively but absolutely smaller than that of a number of small, smooth-brained rodents. With every step in the progress of distortion of the rhinencephalon and the diminution of its primary receptive apparatus, there has taken place a progressive increase in the number of pathways which connect it with the newer parts of the cerebrum. Great numbers of these pathways are deposited beneath the cortex of the cingulate fissure. Its late appearance in the human fetus ceases to be a problem in the light of the fact that the pathways which it contains (and in conformity to which, as we shall have reason to see later, the fissure itself was not improbably formed) are relatively new structures in the history of the phylum.

The gross form of the different types of the cingulate fissure in man has been described by a number of anatomists. Notwithstanding the fact that this fissure is continuous in a large proportion of hemispheres, Retzius (12) agreed with Eberstaller (13) that it can hardly be considered as a single furrow. Both authors considered this fissure as consisting of an anterior part, opposite the anterior bend of the corpus callosum; an intermediate, parallel to the superior part of the corpus callosum; and a posterior, which passes backwards as the posterior boundary of the paracentral lobule.

Of a number of hemispheres investigated by Eberstaller, the fissure was continuous in 68 percent. In the rest, it was double for some distance between the anterior and intermediate parts, the former passing above the latter; in a small number of hemispheres the fissure was severed into three, four or five parts. Retzius found in 100 hemispheres a single continuous fissure in 41; divided into two parts in 44; into three parts in 14; and into four parts in one hemisphere.

The fissure of the hemisphere pictured in Figure 12 was of the continuous type. It was, moreover, united behind with a subparietal (E. Smith) fissure, placed below the precuneus. The pathways of a divided type of the fissure will be mentioned under a separate heading.

The pathways of the cingulate fissure.—These may be classed, according to their different directions, under the headings of transverse, longitudinal and diagonal.

1. Pathways which extend more or less squarely around the fissure, between the outer and inner borders may be seen on the map to cover it throughout its extent, with the exception of its rostral termination. In the situation of some of its tributary sulci, the transverse pathways extend between the particular offshoot and the opposite border of the parent fissure.

2. Longitudinal pathways extend either between the outer or the inner border of the fissure and (a), the opposite border; (b), the postero-superior tributary sulcus; (c), the posterior extremity of the subparietal portion of this cingulate fissure; (d), the antero-superior tributary sulcus.

3. Diagonal pathways: between the superior and the inferior borders of the subparietal termination of the fissure; between some of the offshoots and the opposite border of the parent fissure; between the opposite borders of some of the offshoots; between an inferior offshoot placed on the cingulum convolution (outlined but not mapped on the p cture) and two superior offshoots.

The pathways of a part of the cingulate fissure of the two-piece type.—The variation of the fissure in which the posterior end of the pars anterior passes for some distance above and back of the anterior end of the pars intermedia was examined for the underlying intercortical pathways. The intervening thin convolution was shaved off after the piece was dissected out in the manner described, so that the fiber side of the fissure could be spread out on a glass slide. It was found that although many of the shorter fiber bundles were interrupted in the intervening convolution, at least two sets of longer bundles passed beneath its base in somewhat curved diagonals which crossed each other about the middle of their courses. From the point of view of the intercortical pathways, therefore, the difference between a continuous and an interrupted type of the fissure was in that case small.

3. The Sulcus of the Cuneus (Figures 6 and 15)

The intercortical system of the cuneal sulcus of this hemisphere is illustrative of a simple and basic type of the pattern of these pathways. In the hemisphere pictured this sulcus consisted of three straight limbs which radiated from a central point. Two sets of fiber bundles traversed the fissure, crossing each other in every part.

One set of parallel bundles may be seen to extend in the long axis of the antero-inferior limb and into the superior limb as far as its posterior border. Parallel to these long bundles are short ones which extend almost directly around from border to border of the postero-inferior limb of the fissure. Another set of fibers passes in the long axis of the postero-inferior limb and extends as far as the anterior border of the superior limb. Parallel to these long bundles are short ones which pass transversely around the antero-inferior limb of the fissure.

In this manner the superior limb of the fissure is covered by two sets of long bundles which cross each other; while each of the inferior limbs is covered by the crossing of longitudinal and transverse bundles.

4. The Calcarine and the Parieto-occipital Fissures (Figures 6 and 15)

The angle of the bifurcation in which the posterior calcarine fissure commonly terminates at or near the occipital pole may be from 45 degrees, or less, to 180 degrees. In the latter case the appearance of the terminal sulcus is that of a vertical slit on the occipital pole, three or four centimeters long, near the middle of which it is joined mesially by the posterior calcarine fissure. In a certain number of cases the posterior calcarine is separated from the straight or bent small terminal sulcus by a thin convolution. The latter fact does not militate, however, in favor of a different type of calcarine fissure. It is explained by morphologists on the ground that the small terminal sulcus appears in the fetus as a secondary or tertiary furrow which, at a later stage of the progress of the posterior calcarine fissure towards the occipital pole, becomes, in most instances, united with it; if the backward extension of the posterior calcarine is arrested at a certain stage, it re-

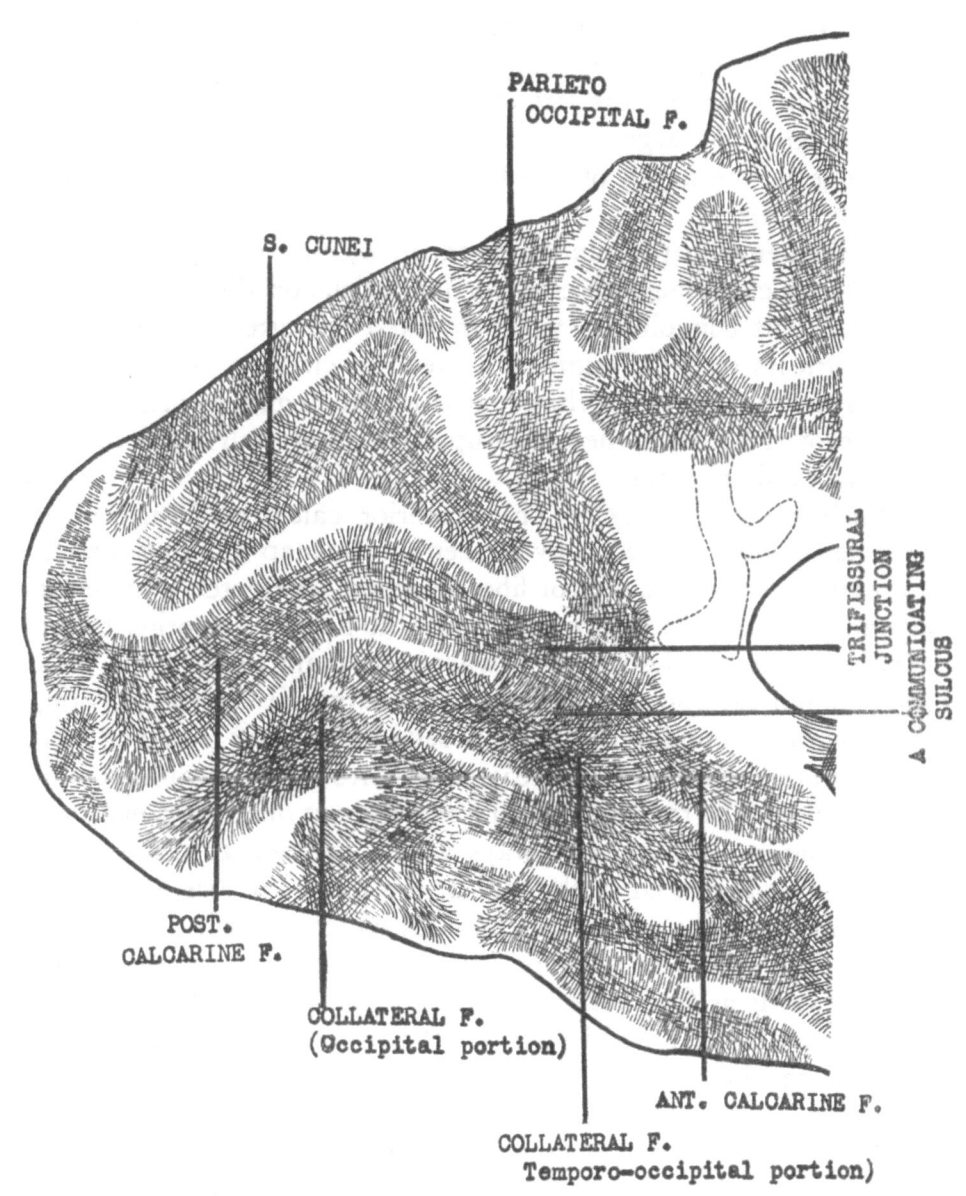

FIG. 15. A COMPOSITE SCHEMA OF THE INTERCORTICAL SYSTEMS OF THE LEFT MESIAL OCCIPITAL AREA OF A HUMAN CEREBRUM

mains separated from its terminal portion. Many variations of the posterior calcarine fissure are thus reducible to a single type.

Yet even in the case of complete coalescence of the two parts of the fissure in question, the site of their union is seldom obliterated, being marked in the large majority of instances by a more or less prominent ridge in the depth of the fissure—the posterior cuneo-lingual annectant gyrus.

In the hemisphere from which the map, Figure 15, was prepared, the posterior calcarine entered a vertical sulcus on the occipital pole. The annectant convolution which marked the site of union of the fissure with its terminal segment, instead of crossing the fissure and bridging the space between the cuneus and the lingual gyrus, extended from the lower border of the fissure upwards about halfway across it in the form of a sharp ridge.

Pathways (Figure 15).—The posterior calcarine fissure of the whole hemisphere studied, may be seen on the map to be traversed in its greater part by two sets of fiber bundles. One extends somewhat diagonally around the fissure from border to border; the other extends in its long axis. In the situation where the main limb of the fissure bends around the polar margin to join its terminal vertical portion, a third set of fiber bundles is superimposed on the other two. The fact is that about one centimeter before its entry into the vertical furrow, the calcarine fissure of that hemisphere had a short offshoot downwards. Some of the intercortical pathways which are seen to traverse that offshoot, may be traced in the roof of the parent fissure, where they are parallel with the other fiber bundles which cross it. Other pathways in the offshoot in question may be observed to curve upwards and backwards as far as the lateral lip of the vertical furrow on the occipital pole, interrupted in part by the posterior annectant.

Near the junction of the posterior calcarine with the parieto-occipital and the anterior calcarine fissures, the longitudinal fiber bundles may be seen on the map to separate into two sets. One set of bundles streams forwards towards the junction of this fissure with the parieto-occipital; the other extends downwards and forwards, towards the anterior calcarine. Several other sets of pathways are superimposed upon these. The significance of the intricate plexus of intercortical fibers in this situation will be more appreciated by a cursory review of the morphologic and embryologic facts involved.

The posterior calcarine fissure extends in a somewhat wavy, generally horizontal course on the mesial surface of the occipital lobe, between the occipital pole and the lower end of the parieto-occipital fissure. Below the latter, the fissure bends forward and downward at an obtuse angle to reach the junction of the posterior portion of the cingulate with the hippocampal convolution. This is the anterior calcarine fissure. For all one can tell from a superficial observation, the latter is as much a downward continuation of the parieto-occipital as a forward continuation of the posterior calcarine. With respect to its depth and the perpendicular structure of its walls, it is indeed more like the former fissure. Moreoever, embryologically the parieto-occipital and the anterior calcarine fissures are developed at the same time, while the posterior calcarine is of later development.

In a small percentage of cases (about 3 percent) of human brains, however, the antero-inferior angle of the cuneus joins the cingulate gyrus, thus completely separating the parieto-occipital from the calcarine fissure. And this rather rare variation of the human cerebrum is, according to Cunningham (14), a constant feature of the cerebrum of the chimpanzee. It is less marked in the orang. "In the ape," says Cunningham, "the stability and great depth of the calcarine fissure shows that it is of vastly greater morphological importance than the parieto-occipital fissure which is extremely variable and often insignificant."

Ecker (15), moreover, has shown that even in the cases in which the cuneus is not united with the cingulate gyrus on the surface, the separation of the parieto-occipital from the calcarine fissure is always marked by a more or less prominent extension of the interior angle of the cuneus in the depth of the fissure. This annectant convolution—the gyrus cunei—is generally in the form of a more or less wavy linear prominence which extends along the bottom of the junction of the three fissures until it touches the posterior border of the cingulate gyrus. "Under no circumstances whatever," maintained Cunningham (14), "do we ever find the gyrus cunei completely absent in the anthropoid brain and thus there never occurs a perfectly free communication between the calcarine and the parieto-occipital fissures."

Although the gyrus cunei is not a serious barrier to the intercortical pathways, a number of them passing beneath its base, we now know that Cunningham's observations on the subject of these fissures have a profounder significance than was thought at the time. The investigations of the last three decades have firmly established the important part played by the area striata of the calcarine fissure as a primary cortical area of vision, to the exclusion of the parieto-occipital fissure.

Cunningham contended that the two portions of the calcarine fissure, the posterior and anterior, are morphologically distinct structures, and that of the two, the anterior is the more fundamental. He adduced in support of his contention the observation that the anterior calcarine appears in the fetus, as a separate furrow, before the posterior. The latter is therefore to be considered as a secondary fissure which is united with the primary one at a later stage of development. In a large number of cases, indeed, the separate origins of the two parts of the fissure remain marked in the adult by an annectant gyrus which extends from the lower part of the cuneus to the lingual gyrus—the anterior cuneo-lingual gyrus.

He further argued that the calcarine fissure in the chimpanzee and the orang is of the same depth throughout its length, with smooth walls and without a trace of any annectant gyrus. He concluded that the entire calcarine fissure of the ape's brain corresponds to the anterior portion in the adult human brain,

and with the entire length of the precursor of the calcarine fissure in the human fetal brain; in the human brain the posterior part of the precursor of the calcarine fissure is obliterated, and in its place a secondary sulcus is laid down at a later date (the posterior calcarine sulcus. In the brain of the ape there is no representative of the latter sulcus.

Retzius (12) did not share Cunningham's view regarding the fundamentally different morphology of the anterior and posterior parts of the calcarine fissure.

In the latter half of the fifth or the beginning of the sixth month of the fetal life, [says Retzius], it sometimes happens, by no means constantly so, that a new furrow makes its appearance posterior to the first calcarine formation, the two gradually uniting to represent the calcarine fissure of the adult brain.... I do not believe, as does Cunningham, that there is a fundamental difference between the anterior and posterior parts of the fissure. The front of the fissure develops at least as often without the posterior addition and extends itself backwards to form the posterior part.

Without going into the merits of the case further, it may be said at once that the intercortical pathways of the mesial portion of the occipital lobe of the adult human cerebrum make for a degree of integration which removes it a vast distance from the corresponding part of the ape's cerebrum and from that of the six months' human fetus. The vestigial remains which mark the phylogenetic and ontogenetic history of the calcarine and parieto-occipital fissures—the gyrus cunei of Ecker and the posterior and anterior annectants—assume a different significance from the point of view of the intercortical pathways. The entire region between the fusiform gyrus below and the precuneus and the posterior border of the cingulum in front, is by these, as well as by other intercortical pathways to be described later, bound up into an anatomical whole.

Pathways of the trifissural junction.—It was pointed out that the longitudinal pathways of the posterior calcarine fissure may be seen on the map (Figure 15) to be separated anteriorly into two contingents. Both enter into (or emerge from) the trifissural junction. One extends upwards and forwards and some of its bundles may be traced on the lowest portion of the anterior border of the parieto-occipital fissure, that is, the posterior border of the cingulate convolution. In their course, some of these pathways pass beneath the base of the gyrus cunei of Ecker; others are interrupted in a mass. The other contingent streams downwards and forwards (or in the reverse direction) below the level of the gyrus cunei and may be traced in the anterior calcarine as far as its lower border and its inferior termination.

The trifissural junction is traversed by still other pathways. In the hemisphere pictured there was a sulcus which established a communication between that junction and the collateral fissure. It will be observed in Figure 15 that by way of this communicating sulcus, pathways extend between the fusiform gyrus on the one hand and the antero-inferior border of the cuneus and the posterior border of the cingulate gyrus on the other.

Pathways of the anterior calcarine fissure (Figure 15).—Three sets of pathways have been mapped in this fissure:

1. Longitudinal pathways extending between the posterior calcarine and the lower end and the inferior border of the anterior calcarine fissure.

The highest of these long bundles curve upwards and are lost to view in the dense plexus of the trifissural junction. Their upward curve, however, is strongly suggestive of the probability that their further course is in the parieto-occipital fissure.

2. Pathways which extend between the superior border of the fusiform gyrus and the lowest portion of the posterior border of the cingulate gyrus.

The course of these pathways is diagonally around the temporo-occipital portion of the collateral fissure, then, by way of the communicating sulcus, diagonally around the posterior half of the anterior calcarine.

3. In the lowest—most anterior—part, fiber bundles extend squarely around the fissure.

Pathways of the parieto-occipital fissure as far as the incisure (*Figure 15*).—Four sets may be seen on the map:

1. Between the upper half of the posterior border of the cingulate gyrus, and the trifissural junction.

2. Long diagonal pathways between points on the posterior border of the fissure (anterior border of the cuneus) and points at much higher levels on the anterior border (precuneus).

In the upper half of the fissure, the fiber bundles curve over the upper border of the hemisphere into the parieto-occipital incisure.

3. A few pathways extend transversely around the fissure about its middle third. These pathways are covered by others which extend around the fissure in short diagonals. At a higher level, the latter assume a transverse course and become therefore parallel to the transverse bundles in the middle third of the fissure.

4. Between the lower half of the anterior border and somewhat higher points on the posterior border.

Pathways of the parieto-occipital incisure (*Figure 22*).—In the hemisphere studied, the part of the parieto-occipital fissure which incised the upper border appeared on the lateral surface in the form of a trifolium. Five sets of pathways have been mapped in it:

1. Continuations of the longitudinal pathways from the fissure on the mesial surface into the central leaf of the trifolium. They, therefore, extend between the superior parieto-occipital region on the lateral surface and the anterior border of the cuneus.

2. Short pathways around the anterior and the posterior leaves of the trifolium.

3. Curved pathways which extend between the upper part of the posterior border of the fissure on the mesial surface and the anterior branch of the trifolium.

4. Short, diagonal pathways in the central and in the posterior leaves.

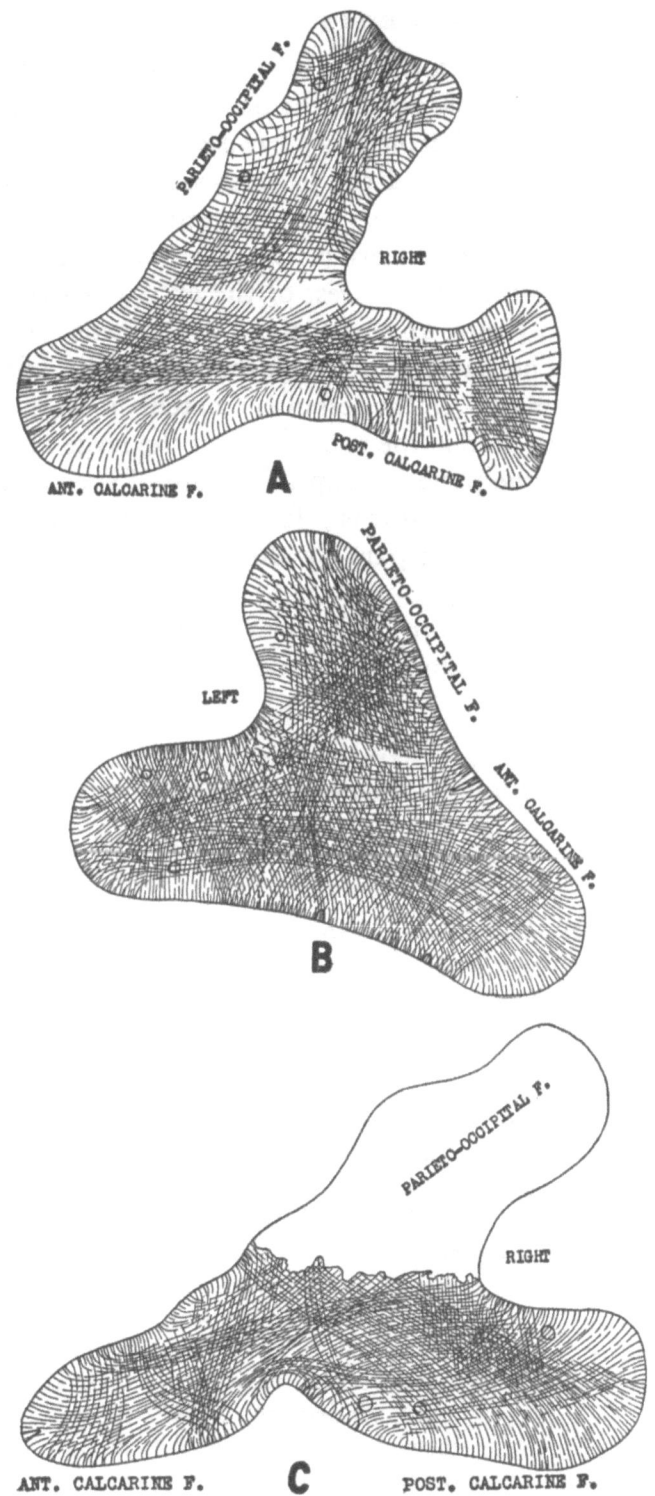

FIG. 16. INTERCORTICAL SYSTEMS OF THE PARIETO-OCCIPITAL AND CALCARINE FISSURES OF THREE HUMAN HEMISPHERES

In spite of apparent differences, an analysis of the courses pursued by the pathways exhibits striking similarities.

5. Pathways in the form of a narrow, slightly curved band with the convexity downwards, around the base of the trifolium.

Other types of the calcarine and parieto-occipital fissure system.—In the type of the calcarine and the parieto-occipital fissure system described, a communicating sulcus which entered the trifissural junction served to complicate the pattern of the pathways in that situation, which was rather simple in the rest of the extent. In the three specimens to be seen in Figure 16, A, B, and C, there was no such communicating sulcus. The pattern of the pathways in the trifissural junction is therefore simpler. In the rest of the extent, it is, however, somewhat more complex than in the type given on the large map. The intercortical lamina of the parieto-occipital fissure of specimen C was lost in the handling; nevertheless it is safe to judge the course of a number of pathways in the absent part by their course in the anterior and posterior calcarine fissures.

Notwithstanding minor differences in the following three types, they are similar in a number of respects, as may be seen from the following:

A TABLE OF SIMILARITIES AND DIFFERENCES OF THE INTERCORTICAL PATHWAYS IN THREE TYPES OF THE CALCARINE AND PARIETO-OCCIPITAL FISSURES

A	B	C
	I	
Between the anterior extremity of the anterior calcarine and the parieto-occipital.	Between the lower border of the anterior calcarine and the parieto-occipital fissures. The course of a number of pathways in the parieto-occipital fissure appears to bend toward the extremity of the anterior calcarine.	Between the extremity and the lower border of the anterior calcarine and the parieto-occipital fissures.
(The most anterior contingents of these pathways are more vertical than the posterior. The two contingents therefore cross each other in the anterior portion of the anterior calcarine.)	(Pathways between the anterior border of the parieto-occipital and the lower border of the anterior calcarine, being more vertical than the other contingents, cross them about the middle of the anterior calcarine.)	(Pathways in the form of a band between the lower part of the anterior border of the parieto-occipital fissure and the lower border of the anterior calcarine, being more vertical than the former contingents, cross them about the middle of the anterior calcarine fissure.)

A Table of Similarities and Differences of the Intercortical Pathways in Three Types of the Calcarine and Parieto-occipital Fissures (Continued)

A	B	C
	II	
Between the posterior extremity of the posterior calcarine and the anterior extremity, as well as the upper border of the anterior calcarine.	Between the posterior extremity of the posterior calcarine and the extremity, as well as the adjoining parts of both borders of the anterior calcarine.	Between a part of the extremity, as well as the adjoining border of the posterior calcarine and the posterior half of the inferior border of the anterior calcarine.
	III	
Diagonal pathways between the anterior part of the superior border of the posterior calcarine and the inferior border of the anterior calcarine.	Diagonal pathways between the anterior part of the superior border of the posterior calcarine and the inferior border of the anterior calcarine.	Diagonal pathways between the anterior part of the superior border of the posterior calcarine and the inferior border of the anterior calcarine.
	IV	
Squarely or diagonally around the posterior calcarine.	Squarely or diagonally around the posterior, as well as the anterior calcarine	Squarely or diagonally around the anterior and partly around the posterior calcarine.
	V	
Between the anterior border of the parieto-occipital and the incisura.	Between the anterior border of the parieto-occipital and the incisura.	
	VI	
Squarely or diagonally around the parieto-occipital.	Squarely or diagonally around the parieto-occipital.	

Other minor likenesses and differences may easily be gathered from the maps.

5. Deep Intercortical Systems of the Occipital Lobe

Immediately lateral to the intercortical pathways of the fissures described in the foregoing, there extends a sheet whose fibers belong to the class usually known as "long association fibers." It is of a generally triangular outline, like that of the cuneus, but larger than the latter by the widths of the parieto-occipital and posterior calcarine fissures and the length of the anterior calcarine fissure. Its thickness is less than 0.5 millimeter. It is vertically placed, its plane being parallel with the mesial aspect of the occipital lobe. Its surface is deeply indented by a number of irregular curves. A fairly correct idea of its general appearance may be gained from the following: As one faces the right occipital pole, a coronal section through the vertical sheet passing through the parieto-occipital and posterior calcarine fissures would appear in the form of a line bent in three curves; the upper and the lower curves with the convexities to the right, corresponding to the outlines of the two fissures; the middle curve with the convexity to the left corresponding to the prominence of the cuneus. If a coronal section were made through the anterior calcarine fissure, then the sheet in question would appear in the form of a linear curve with the convexity to the right, corresponding to the outline of the anterior calcarine fissure lateral to which it is placed.

Vertical and vertico-transverse intercortical systems have been described in the general region dealt with by a number of reliable investigators; they have been doubted, disputed or denied, in whole or in part, by equally reliable investigators. A mere mention of some of the facts involved will suffice.

Wernicke (16), (17), has described a bundle of fibers in the cerebrum of a monkey, which he thought connected the fusiform gyrus with the upper part of the inferior parietal lobule—the fasciculus occipitalis verticalis or perpendicularis. He located the bundle lateral to the "sagittal marrow." If by the latter term is understood the combined layer of fibers of the optic radiation and the corpus callosum which lie in the roof, the outer wall and the floor of the posterior horn of the lateral ventricle, then the bundle in question, on its way from the fusiform gyrus to the inferior parietal lobule, must at first proceed horizontally beneath the floor of the ventricle, then curve forward around the outer wall of that cavity. It is therefore difficult to see how the direction of the bundle could be vertical. The existence of such a bundle in whole or in part, in man, was thought probable by Sachs (18), Dejerine (19), and others. Probst (20) saw a bundle in the general region pointed out by Wernicke, but could not say whether it was an intercortical bundle or not. Schnopfhagen (21) denied its existence, and maintained that Wernicke had mistaken certain projection and callosum fibers for an intercortical pathway. Mayendorf (1) thought he saw that bundle in the region indicated in coronal sections (in the same manner that Probst did), but he could not determine its origin and termination. In a Weigert-Pal preparation of a cerebrum in which the callosal and the projection fibers were in that region degenerated, while the fusiform gyrus was intact, he could not find it. His conclusion was that there is nothing to prove its existence.

The question regarding Wernicke's bundle becomes more complicated by a description of another fiber tract by Vialet (22). It originates, according to the author, in the lingual gyrus and in the calcarine fissure, and is placed immediately lateral to the inferior longitudinal fasciculus and parallel with it, being distinguishable from the latter by the finer caliber of the fibers and by their paler Weigert-Pal stain. The bundle bends around the ventricle and radiates on to the convexity of the occipital lobe. Probst (23) thought that the bundle in question was part of the "sagittal marrow." Mayendorf (1) was of the same opinion.

The degree of uncertainty with which the question of these fiber tracts is pervaded, makes it necessary to describe in some detail the technic of the method by which the vertical triangular sheet, investigated in the present study, was obtained.

Technic.—This applies to the manner of obtaining the fiber sheet in question, as well as to the other intercortical systems of the mesial portion of the human hemisphere. A valuable tool for this purpose is a chisel about 12 centimeters long, slightly curved in its long axis, made of some plastic material such as hard rubber or viscose. The one I have used has a rounded edge about one centimeter wide, which is not sharp enough to cut the tissue. This tool is inserted through the septum lucidum into the lateral ventricle and pushed forwards, upwards, downwards and backwards parallel to the lateral wall of the cavity. At first the aim is to cut, or rather to tear, through the lower layers of the callosal band—through those of its fibers which perforate horizontally the corona radiata, as well those which descend on the posterior part of the ventricular wall as the tapetum. Beyond this the chisel is guided along the more or less natural plane of cleavage between the fibers of the corpus callosum and those of the cingulum which radiate towards the upper border of the hemisphere. Posteriorly the chisel will in all

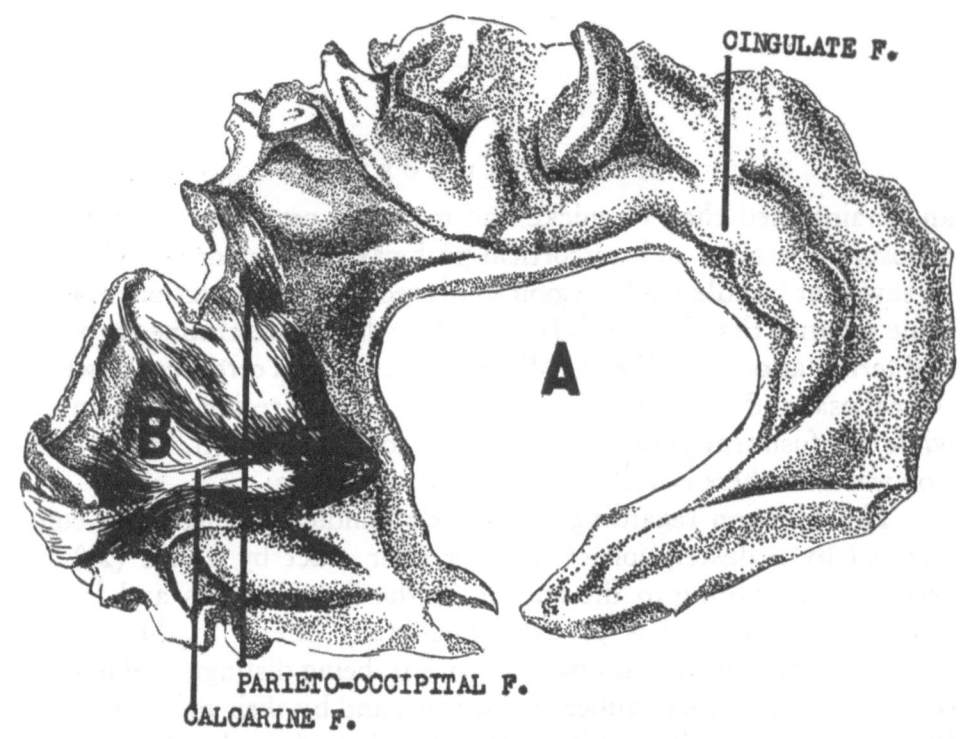
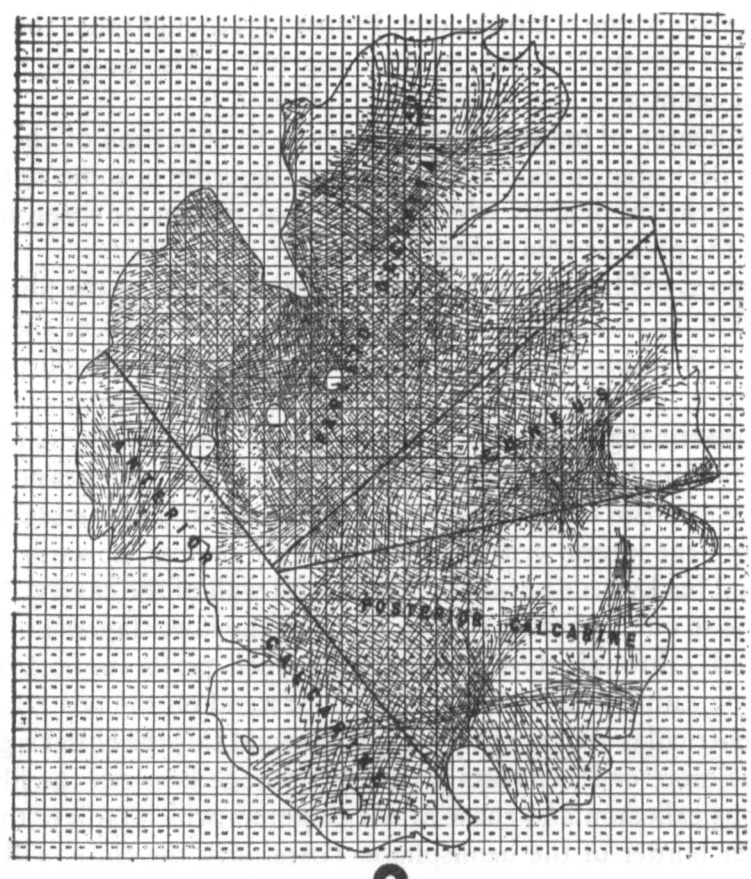

FIG. 17. THE GROSS AND MICROSCOPIC APPEARANCES OF THE DEEP INTERCORTICAL SYSTEMS OF THE OCCIPITAL LOBE

A, the white laminae underlying the cortex of the fissures of the mesial surface of the right hemisphere of a human cerebrum. The drawing was made from a metal cast of a dissection of an exploded brain. At B the dissection is incomplete, showing the triangular deep fiber sheet whose microscopic appearance may be studied in C.

probability pass along the fibers of the optic radiation and of the forceps major of the callosum. In the temporal cavity the tool is aimed at the rhinal fissure; in the frontal, at the gyrus rectus and the anterior border of the hemisphere.

The large piece of tissue which is thus separated is rough on its lateral aspect, being covered by shreds of callosal fibers for the greater part of its extent and by shreds of the optic radiation and callosal fibers posteriorly. The cut edge of the corpus callosum remains adherent to the inner oval of the cingulum. By pulling gently upon the cut surface of the callosal band, the lateral aspect of the preparation is cleaned off with comparative ease. The exposed surface consists of the fine radiating fibers of the cingulum as far as the parieto-occipital and anterior calcarine fissure. Beyond that, as far down as the anterior and posterior calcarine, the surface consists of the smooth triangular sheet in question (Figure 17 B).

If the dissection is carried slightly deeper, the structures exposed throughout consist of the hull-like prominences corresponding to the fissures of the mesial aspect of the hemisphere (Figure 18). Working with a hemisphere exploded by the method described, the entire process is much facilitated.

The preparation of microscopic sections from the triangular sheet thus exposed by dissection is accomplished by the method already described. It need only be added that the thinness of the sheet, together with the fact that many of its constitutent fibers pass in or out of the cortex throughout its extent by perforating the mesially-lying "short" intercortical structures and the thickness of the cuneus, make it inadvisable to attempt to dissect it away from the other mesial structures to which it is adherent. The triangular piece is simply cut out, the prominence of the cuneus and the cortical gray substance are shaved off, and the resultant more or less flat piece of tissue spread out with the lateral surface applied to the glass. The first ten or twelve sections (30 to 50 micra) will pass through the plane of the sheet.

Deep intercortical pathways of the mesial occipital area.—These are shown in Figure 17 C. It is a drawing of a reconstructed section. The irregular outline is due to the fact that the sheet, being very much indented and bent in several directions, was cut on the edges radially, as described, before it could be flattened on glass. The small notches on the edges and the circles near the edges are the markings made on the flat block for the purpose of orientation.

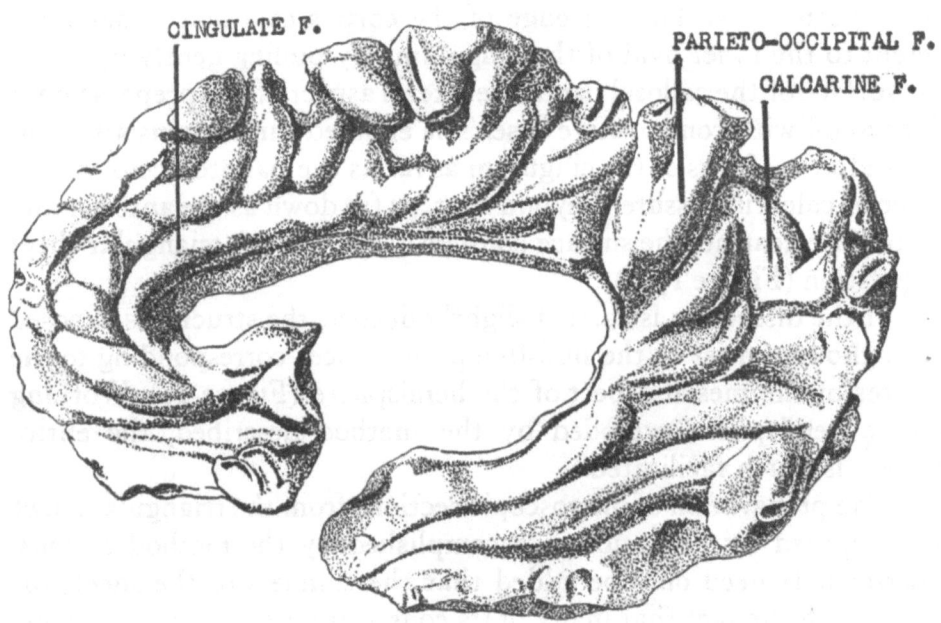

FIG. 18. THE WHITE LAMINAE UNDERLYING THE CORTEX OF THE FISSURES OF THE MESIAL SURFACE OF THE LEFT HEMISPHERE OF A HUMAN CEREBRUM

An obverse of the configuration of the cortex. The drawing was made from a plaster-of-Paris cast.

The following are the main pathways:

1. Between the entire extent of the cuneus and (a), the parieto-occipital fissure; (b), the posterior calcarine fissure; (c), the lower wall and floor of the anterior calcarine fissure.

2. Between the parieto-occipital fissure and (a), the posterior calcarine fissure; (b), the superior wall and floor of the anterior calcarine fissure.

Connections are therefore established in a number of different combinations, between the posterior parts of the precuneus, the cingulate and the hippocampal convolutions, the gyrus lingualis and the superior margin of the hemisphere between the lingual gyrus and the precuneus.

CHAPTER SEVEN

INTERCORTICAL SYSTEMS OF THE BASAL TEMPORO-OCCIPITAL AREA

1. THE COLLATERAL FISSURE (FIGURES 15, 19 AND 20)

The anterior portion of this fissure bounds the outer border of the hippocampal convolution in man, and the pyriform area throughout a large series of the lowest mammalia. It may therefore be considered as one of the most ancient in the history of cerebral fissuration. About the middle of its course, the infolding of the entire thickness of the cerebral wall in the fetus is manifested in the adult by the eminentia collateralis in that part of the ventricular cavity. But the extension of the collateral fissure on the occipital lobe cannot be older than that lobe itself. Moreover, the pathways contained in the hippocampal portion of this fissure must be of relatively recent standing, since they can hardly have any other meaning than as connecting links between an ancient and relatively new cerebral structures. In the light of these considerations, the late appearance of the collateral fissure in the human fetus—in the beginning of the sixth month—like the late appearance of the cingulate fissure, becomes understandable.

The pathways contained in this long fissure establish so many connections in the several regions which it traverses that in the interests of lucidity it may be divided into three portions, an occipital, a temporo-occipital and a hippocampal. The temporo-occipital portion was in this hemisphere separated from the anterior and the posterior portions by annectant convolutions.

Pathways of the posterior or occipital portion.—1. Almost vertically directed pathways (as they appear on the maps, Figures 15 and 20) between the upper and lower borders of the fissure.

2. Pathways in the long axis of this portion of the fissure.

3. Almost horizontally directed pathways between its superior border and the superior border of a sulcus which furrows the fusiform gyrus, parallel to and below the temporo-occipital portion of the collateral fissure.

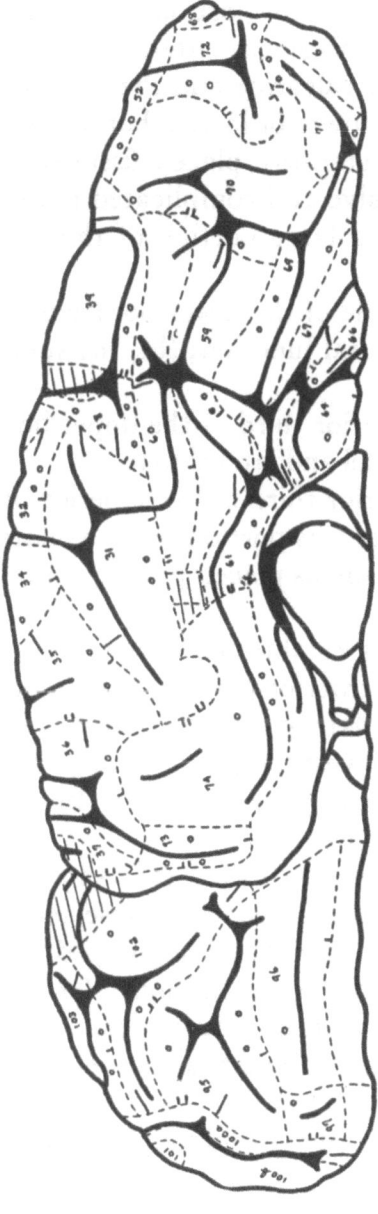

FIG. 19. A LINE DRAWING OF THE BASAL SURFACE OF THE PLASTER-OF-PARIS CAST OF THE LEFT HEMISPHERE OF THE CEREBRUM (SEE LEGEND, FIGURE 6)

Pathways of the temporo-occipital portion.—The pathways of this portion of the collateral fissure are complicated by a narrow sulcus which in this hemisphere passes transversely across the temporo-occipital region, severing the anterior portions of the lingual and fusiform gyri. Above and mesially this sulcus entered the trifissural junction of the anterior and posterior calcarine and the parieto-occipital fissures. In its course towards the lower or lateral border of the hemisphere, it crossed all the antero-posteriorly placed fissures of that region. By way of this sulcus, extensive and complicated pathways may be seen on the maps to traverse a large portion of this region of the hemisphere:

1. Between the middle of the lower border of this portion of the fissure and (a), the trifissural junction; (b), the superior border of the anterior calcarine fissure.

2. Between the posterior third of the upper border of this fissure and the sulci placed parallel and lateral to it.

3. Between the anterior third of the superior border and (a), the annectant which separates this portion of the collateral from the hippocampal portion; (b), a sulcus placed parallel and lateral to this portion of the collateral.

Connections are thus established between the lingual and fusiform gyri and the antero-inferior parts of the cuneus, the lower posterior portion of the cingulate gyrus, the posterior portion of the hippocampal gyrus and the lowest temporo-occipital gyrus.

Pathways of the hippocampal portion.—1. Between the annectant and the anterior extremity.

2. More or less diagonal sets of pathways around the fissure. It will be seen from the map that none of these sets covers the fissure throughout its extent and that they are grouped in such a manner as to cross each other in only three situations: for a short space about the middle of the fissure and at its extremities.

3. Between points at different distances apart on the lateral border.

4. Between points about the middle of the lateral border and the anterior extremity.

2. The Inferior Longitudinal Fasciculus

Like the arcuate, the inferior longitudinal fasciculus has been a source of contention for many years. On the evidence of gross dissec-

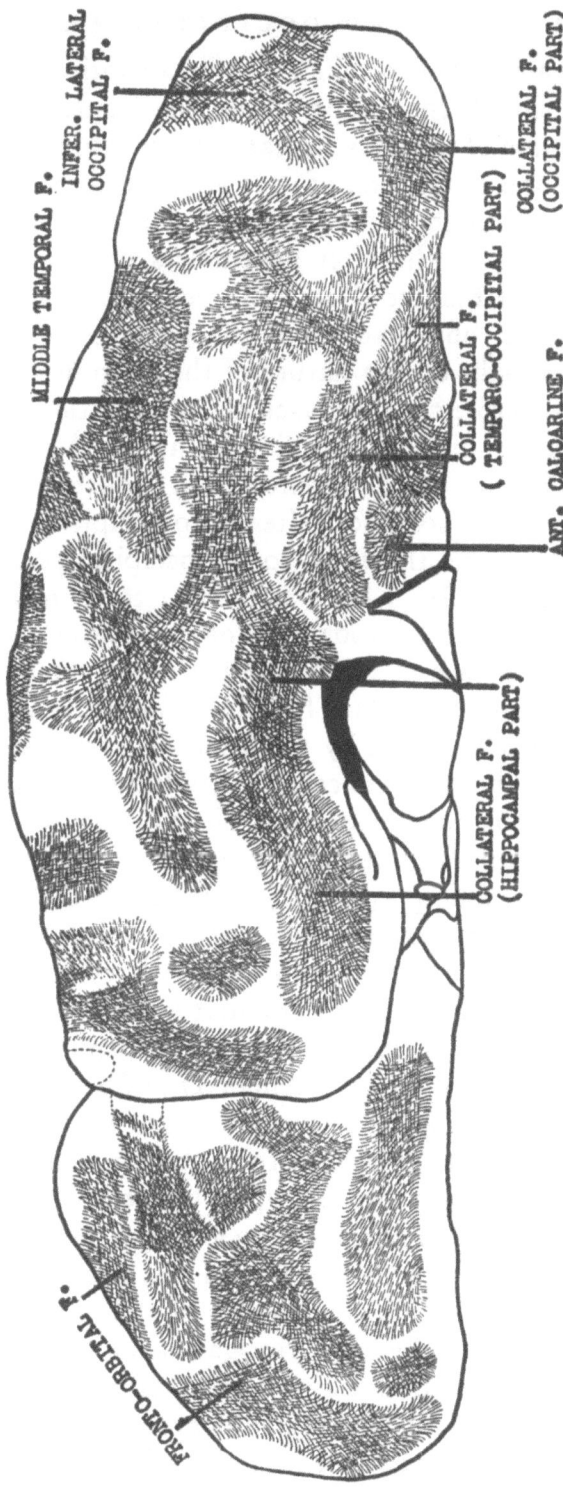

FIG. 20. A COMPOSITE SCHEMA OF THE INTERCORTICAL SYSTEMS OF THE LEFT BASAL AREA OF A HUMAN CEREBRUM

tions, the older anatomists had taken it to be a continuous fiber tract connecting the occipital, the temporal and the frontal poles. By the repeated reproduction of the old schemata of this fasciculus from textbook to textbook, the conception of it as a long continuous tract has become firmly rooted in the minds of many clinicians, but little disturbed by the doubts of the later anatomists on the subject. The advent of the Weigert-Pal and Marchi degeneration methods and of Flechsig's and others' observations of the progress of myelination, have thrown a new light on the nature of the deep intercortical systems. The suspicion arose that the fasciculus in question was a part of the thalamo-cortical radiation. Working with the Marchi as well as with the Weigert-Pal method of degeneration, Probst (20), (23), (24), traced, in the "inferior-posterior" (temporo-occipital) longitudinal fasciculus, fibers from the thalamus to the occipital pole. His statement however, is guarded: "I have arrived at the conclusion," says he, "that we have in the inferior longitudinal fasciculus *mainly* a thalamic radiation." In a case in which the temporal convolutions had been largely destroyed, Quensel (6) found the fasciculus intact, except where it had been itself injured by the lesion. He concluded that the fasciculus is part of the geniculo-calcarine radiation. Anton and Zingerle (25) had previously made a similar observation. Mayendorf (1), (26), on the strength of Weigert-Pal degeneration specimens and that of Flechsig's myelogenetic preparations, arrived at the same conclusion. The opinion formed regarding the nature of the inferior longitudinal fasciculus, after its investigation in the present study, differs from that of both the old and the recent anatomists.

The visual radiation is separated from the ependyma of the temporo-occipital ventricle by the very thin layer of fibers of the tapetum of the corpus callosum. The lower contingents of the visual radiation pass from the geniculate eminence on the pulvinar of the thalamus at first downward and forward into the tip of the temporal cavity, then curve backwards in the floor of the ventricle where they continue their course towards the cortex of the lower wall and floor of the calcarine fissure. The curve of these fibers in the tip of the temporal ventricle has been firmly established by the investigations of Meyer (27), Archambault (28), Cushing (29) and a number of others. In microscopic sections of flattened dissections of this fiber system, I obtained a complete picture of this curve in each of a number of sections (7).

The thickness of the tissue between the tip of the ventricle and the temporal pole is about two or three centimeters. Of this the curved fibers of the visual radiation take up hardly a millimeter. If the curved fibers be removed by dissection, a straight fiber bundle is exposed which proceeds beyond the former to the temporal pole, where it is situated beneath the uncinate fibers with which it is there in part interlaced. If the continuation of the curved fibers be removed by dissection in the floor of the temporo-occipital ventricle, the straight bundle which lies beneath appears to view. Microscopic sections of flattened dissections of the two fiber systems, stained by the method of Weigert-Pal, differ in appearance. The bundles of the visual radiation are, between the anterior curve and the calcarine fissure, straight and long; those of the fasciculus which lies beneath them, short and rather twisted. After the removal of the latter, the long hulls of the boatlike subcortical laminae of the fissures at the base are exposed to view.

On the basis of the foregoing, I am inclined to think that the inferior longitudinal fasciculus is an intercortical system of the same nature as the arcuate (*vide infra*), consisting, like the latter, of fibers of different lengths, the longest of which however, are much shorter than might be judged from the gross appearance of that bundle. The reason why it is not seen to be degenerated in lesions of the temporal convolutions, is, in all probability, because of the relative shortness of its fibers, and because, like those of the arcuate fasciculus, their origin is not in the cortex of the crests of the convolutions but in that of the deeper portions of the fissures. An experimental or other form of lesion deep enough to penetrate that fasciculus would easily involve the optic radiation; a long tract of degeneration must then take place in the course of the latter fibers, which could be traced with comparative ease; but the tract of degeneration must rapidly fade out and disappear in the short and twisted fibers of the inferior longitudinal fasciculus.

3. General Considerations regarding the Cingulate and Collateral Fissures

In his classic monograph on the limbic lobe of the lower mammalia, Broca (30) described the fissure which bounds it as consisting of two arcs, a superior and an inferior one:

The superior arc is placed on the internal surface of the hemisphere, parallel to the convex outline of the corpus callosum, between it and the superior border of the hemisphere. The inferior arc begins in front of the external border of the olfactory lobe, extends along the entire border of the base of the cerebrum, and terminates on the inferior, tentorial, surface of the hemisphere in a part represented in primates by the occipital lobe. The two arcs of this fissure are not, however, joined as are the two arcs of the limbic lobe. They do not surround the lobe completely because the two extremities of the convolution of the corpus callosum are merged in front with the frontal and behind with the parieto-occipital, or rather with the posterior part of the parietal lobe, the occipital lobe not being distinct in animals below primates.

In the human hemisphere pictured in Figures 6, 12 and 15, the superior arc of Broca's limbic fissure is represented by the cingulate fissure; the inferior, by the anterior portion of the collateral fissure. In front, where the deep incision of a portion of the horizontal fissure separates the hippocampal from the rostral convolutions, the gap between the two arcs is apparently complete, unless it be bridged by the uncinate fasciculus. Posteriorly the anterior or hippocampal portion of the collateral fissure was continuous on the surface with its temporo-occipital portion, from which it was, however, separated in the depth of the fissure by an annectant convolution. Some of the pathways contained in the hippocampal portion of the fissure, however, undoubtedly continued their course beyond the annectant, in the temporo-occipital portion.

The collateral fissure was separated, in this hemisphere, from the posterior extremity of a subparietal fissure (which was continuous with the cingulate) by a portion of the lingual gyrus, by the anterior calcarine and the parieto-occipital fissures and by a thin convolution which intervened between the latter fissure and the subparietal. A scrutiny of the intercortical systems of the fissures of that region, however, suggests a much narrower posterior separation of the two arcs of the general limbic fissure. Pathways may be seen to stream between the temporo-occipital portion of the collateral fissure and the calcarine and parieto-occipital fissures. What part these connecting pathways between the visual and rhinal areas may play in the physiological association of the corresponding functions is for the clinician to decide. Instances of associated visual and olfactory hallucinations, such as have been reported by Cushing (29), elaborated on by a detailed study of the pathological anatomy of the intercortical pathways, may throw light on the obscure physiology of these structures.

CHAPTER EIGHT

INTERCORTICAL SYSTEMS OF THE PARIETAL AREA
(FIGURES 21 AND 22)

1. The Interparietal Fissure

Its great length, its numerous variations and the difficulty of determining the homology of certain of its parts have made this fissure one of outstanding interest in morphological literature. As a morphological unit, it was first described by Turner (31) and later by Pansch (32). In the course of the last forty years, a number of anatomists have partly revised and partly repeated the original descriptions. As it stands today, this fissure is usually described as an arch, with the convexity upwards, which traverses the parietal lobe in a sagittal direction. It consists of three principal parts; the first is formed by the inferior segment of the postcentral fissure; the second is the sagittal or horizontal ramus of the arch, or the interparietal fissure proper; the third is the posterior ramus, consisting of a horizontal and transverse or vertical, rather short fissure, the sulcus occipitalis transversus of Ecker (15).

An occasional operculum may be observed on the posterior lip of the latter sulcus, and as the situation of the furrow corresponds more or less to that of a prominent fissure in the same region in apes—the Affenspalte—it has been thought by a number of anatomists to be homologous with the latter. Wernicke (33) was of this opinion. Eberstaller (34) thought the sulcus occipitalis transversus to be merely analogous with the Affenspalte. From Cunningham's (14) writings on this subject, it is obvious that he encountered considerable difficulty in arriving at a conclusion. It seems that the homologue of the Affenspalte, which is generally present in the young human fetus, usually disappears at a later stage. Occasionally, however, it persists and is placed in the adult immediately behind the sulcus occipitalis transversus of Ecker. It is apparently on these considerations that Cunningham says: "I am ... very far from denying that the 'Affenspalte' is invariably absent in man. I only contend that it is not identical with

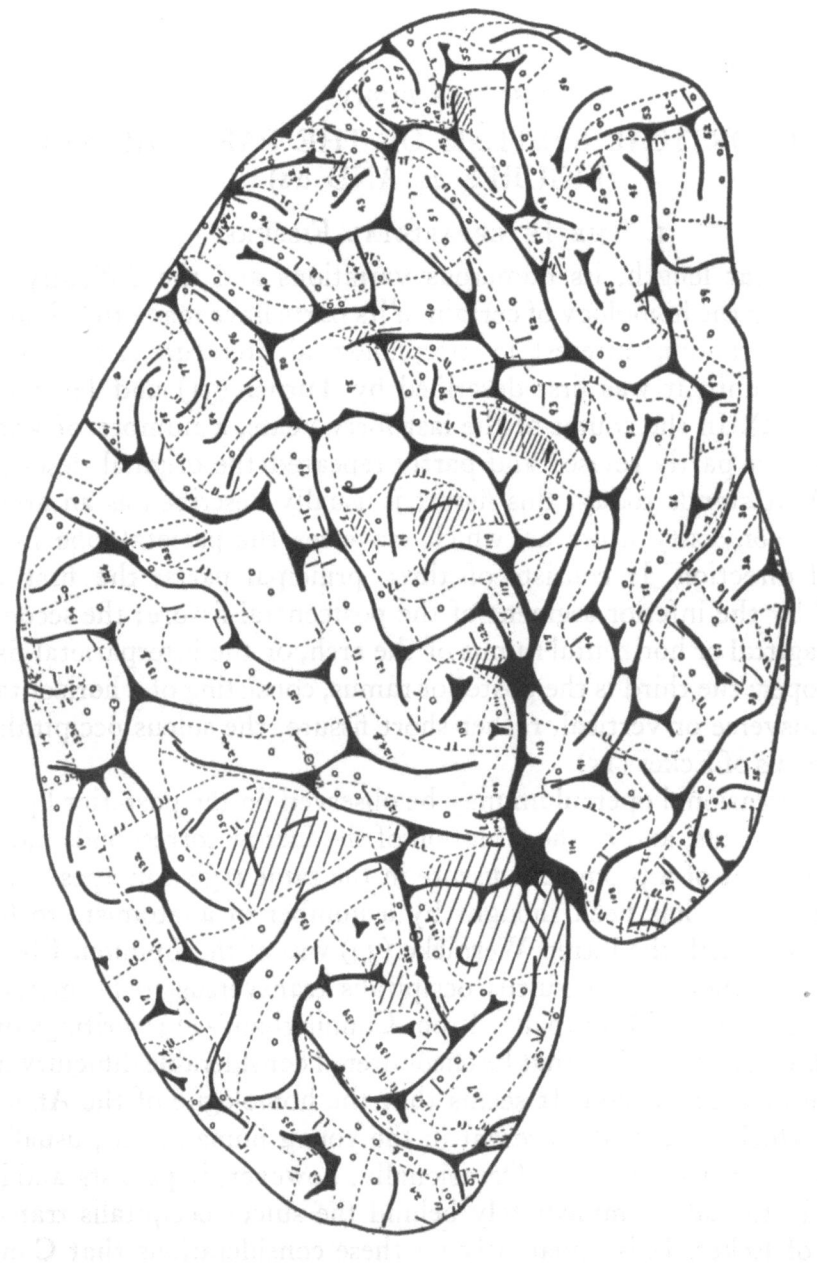

FIG. 21. A LINE DRAWING OF THE LATERAL SURFACE OF THE PLASTER-OF-PARIS CAST OF THE LEFT HEMISPHERE OF THE CEREBRUM (SEE LEGEND, FIGURE 6)

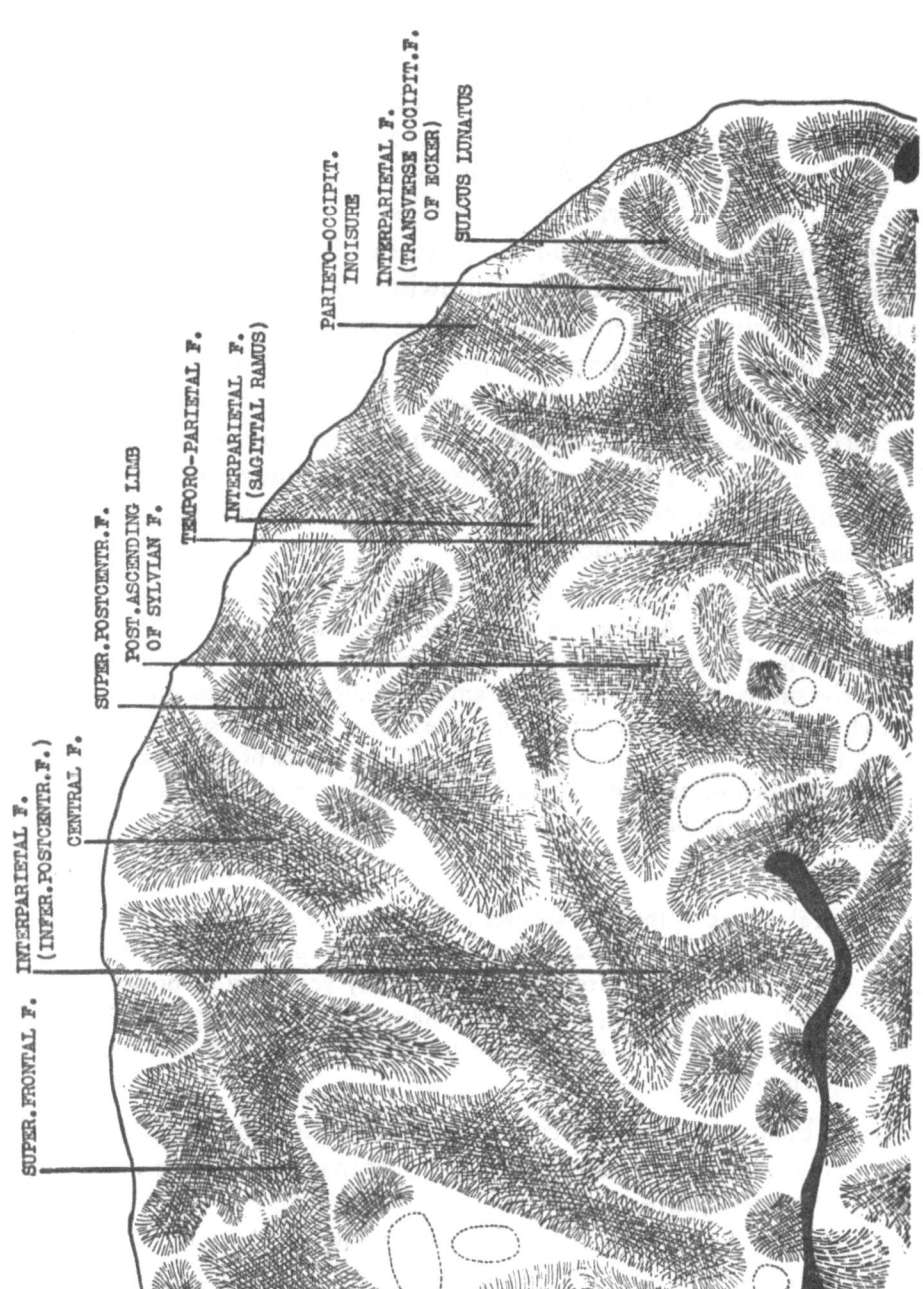

FIG. 22. A COMPOSITE SCHEMA OF THE INTERCORTICAL SYSTEMS OF THE LEFT CENTRAL AND PARIETAL REGIONS OF A HUMAN CEREBRUM

the sulcus transversus of Ecker." His opinion on the subject becomes firmer as he goes along. "The entire sulcus transversus of Ecker," says he on one of the following pages, "is quite independent of the 'Affenspalte' of the apes and has, in point of fact, nothing to do with it." His conclusion is definite: "The sulcus occipitalis transversus of Ecker is not the homologue of the 'Affenspalte' in the apes, but merely a terminal bifurcation of the ramus occipitalis."

Regarding the operculum on the posterior border of the sulcus in question, Retzius (12) offered the following reasonable explanation. The calcarine fissure frequently ends in a vertical furrow on the occipital pole. When the latter happens to be very prominent, the structures adjoining it on the lateral surface are pushed forward in concentric formations of one or two gyri, the latter having a lip which overhangs the fissure. I have encountered this formation frequently. Elliot Smith's opinion on the subject is difficult to interpret. It is best judged from the following quotations:

It often happens [says he (35)], (especially in the brains of lowly human races, such as Negroes and aboriginal Australians and in the anthropoid apes) that the sulcus occipitalis anterior together with the sulcus occipitalis inferior, form a large arc (parallel to the sulcus lunatus), forming the anterior limit of a great tongue of cortex the tip of which often touches the upper end of the sulcus temporalis superior in those cases where there is no sulcus temporo-parietalis. The presence of this great arcuate sulcus explains much of the misleading literature relating to the search for an Affenspalte in the human brain.

And again (36):

Upon the lateral aspect of the hemisphere, in most of the apes, there is a furrow which was supposed to be so peculiarly distinctive of these primates, that it was labelled the *Affenspalte* or ape-fissure. More than twenty years ago its presence was demonstrated in the human brain, and as its old name was clearly inappropriate, the new designation *sulcus lunatus*, in reference to the semilunar form it usually assumed, was given to it.

On the basis of cytoarchitectonic studies, Economo (37) arrived at the conclusion that the sulcus occipitalis transversus represents, on the whole, the Affenspalte.

It is interesting to note that the upper part of the sulcus occipitalis transversus of the hemisphere studied is double. From the foregoing arguments of the morphologists, one is inclined to conclude that the posterior portion of the sulcus represents the Affenspalte while the anterior is the terminal bifurcation of the occipital ramus of the interparietal fissure.

The following table, taken from Jeffersons's (38) morphological study of the interparietal fissure, furnishes a good idea of its variations. It will be seen that the fissure of the whole hemisphere mapped in the present study corresponds on the surface to Type 4, but in its depth, to Type 1.

Type	Description	Percentage		
		Cunningham	Retzius	Jefferson
1	All three parts of the sulcus separate...	6.3	9	35
2	Ramus horizontalis confluent with sulcus postcentralis inferior; postcentralis superior separate.................	19.1	11	26.3
3	Postcentral sulci confluent; ramus horizontalis separate..................	11.1	17	25
4	The three parts of the sulcus confluent.	60.3	55	8.8
5	Ramus horizontalis apparently joined to lower end of upper part of postcentral sulcus; lower end of postcentral sulcus separate........................	2	4	5

The Inferior Postcentral Fissure (Figures 21 and 22)

Orientation.—In the hemisphere pictured, the inferior postcentral was on the surface continuous above with the superior segment of the postcentral fissure and with the anterior end of the sagittal ramus of the interparietal. It was partly separated from both fissures by annectants in its depth. The lower end of this segment of the interparietal fissure incised deeply the superior operculum of the horizontal fissure and terminated on it about halfway between the exposed portion of the operculum and insula. The shape of the inferior postcentral of this hemisphere was that of a right angle, whose point faced foward in the form of a small offshoot of the fissure. The two limbs of this right angle were almost of equal length and the lower one, that which entered the horizontal fissure, described a rather steep anteriorly concave curve.

Pathways of the inferior postcentral fissure.—1. Pathways extending between the convexity of the lower portion of the posterior border of the fissure (the lower limb of the right angle), and the superior border of the anterior offshoot.

2. Between the lower border of the anterior offshoot and the pos-

tero-superior annectant which separates this fissure from the arch of the interparietal fissure. The direction of these pathways may be seen on the map to be continued into the arch of the interparietal in common with certain transversely directed pathways of the fourth set.

3. Between points on the anterior border. The more posteriorly situated contingents of these pathways extend between points successively farther apart.

4. Pathways directed more or less squarely around the fissure throughout its length. In the situation of the anterior offshoot, these extend between that offshoot and the posterior border of the fissure. In the situation of the postero-superior annectant, they extend between that annectant and the anterior border of the fissure. In the latter situation, the course of these bundles may be seen on the map to be continued into the arch of the interparietal fissure.

5. Between points above and below the convexity of the lower part of the posterior border.

6. Between the upper half of the posterior border of the fissure and the superior annectant which separates this fissure from the superior postcentral.

The Sagittal Ramus (Figure 22)

In the specimen studied, this part of the fissure was surmounted by three offshoots. One of these, which in this hemisphere was in the shape of a trifolium, is nearly a constant feature. Below, opposite the antero-superior offshoot, the sagittal ramus communicated on the surface with a supramarginal sulcus, which was a continuation of the posterior ascending limb of the Sylvian fissure. It was partly separated from this sulcus by an annectant convolution.

The differences in the conceptions of morphologists regarding the significance of the annectant convolutions is well illustrated in the case of the fissure in question. Wernicke (33) held that the annectants in the course of this fissure are characteristic of the human cerebrum. Mihalkovics (39) states that prominent annectants in the course of this fissure are indicative of a high degree of cerebral development. Rudinger (40) noted the fact that the sagittal ramus of the interparietal fissure of the primate cerebrum does not contain any annectants. Cunningham (14), who throughout assumes annectants to be a mark of incomplete development, found annectants in this fissure in the brain of the chimpanzee and the baboon.

In the hemisphere studied, the sagittal ramus of the fissure in question contained three annectants. Two of these were merely prominent buttresses on its superior wall.

Pathways of the sagittal ramus.—1. Between the annectant which separates this fissure from the inferior postcentral and (a), the upper border of the anterior leaf of the trifoliate offshoot; (b), the posterior leaf of the trifolium; (c), the extremity of the postero-superior offshoot.

Below these pathways, in the posterior position of the fissure, fiber bundles may be seen in the map which are at first parallel to the former and which gradually assume an increasingly vertical course until, in the most posterior portion of the main fissure, they are nearly vertical and parallel to the posterior annectant which separates this portion of the interparietal from the occipital ramus.

2. Between the extremity of the antero-superior offshoot and the posterior annectant. Above and below these pathways, fiber bundles may be seen extending between correspondingly opposite points on the borders of the fissure, extending into the postero-superior offshoot.

3. Between the lower border of the anterior leaf of the trifolium and the space on the border between the middle and posterior leaves.

4. Between the posterior border of the trifoliate offshoot and its central leaf.

5. Pathways squarely around the central leaf of the trifolium; around the posterior leaf.

6. Between the superior border of the antero-superior offshoot and the annectant which separates this fissure from the supramarginal sulcus. Parallel with these pathways, fiber bundles extend in a somewhat diagonal course around the most anterior part of the fissure as well as around the antero-superior offshoot. Diagonal pathways cross the latter bundles in the anterior offshoot.

The Transverse Occipital Fissure (Figure 22) and the Sulcus Lunatus

Orientation.—Together with its sagittal ramus, the sulcus occipitalis transversus usually describes the form of a horseshoe, in the opening of which is placed the parieto-occipital incisure. In the hemisphere studied, its form was rather that of two horseshoes joined at their convexities along the line of the sagittal ramus. In the enumeration of the pathways it will therefore be more convenient to omit the term sulcus occipitalis transversus and to signify the four limbs of the two horseshoes as offshoots of the sagittal ramus. The sulcus lunatus was in this hemisphere united below with the postero-inferior offshoot, but was separated from the postero-superior offshoot.

Pathways.—1. Between the anterior annectant which separates this fissure from the interparietal proper and the middle third of the lower border. Above these pathways and parallel with them, fiber bundles may be seen extending diagonally across the antero-superior offshoot.

2. Between the middle third of the posterior border and (a), the annectant which separates the antero-inferior offshoot from the temporo-parietal fissure; (b), the posterior border of the antero-inferior offshoot.

Above these pathways and parallel with them, fiber bundles may be seen extending between the anterior annectant and the upper border of the fissure. Still higher, fiber bundles extend parallel with the latter across the antero-superior offshoot.

3. Between the lower border of the fissure and: (a), the antero-superior offshoot; (b), the superior border of the fissure: (c), the anterior border as well as part of the extremity of the postero-superior offshoot.

4. Between the extremity and the posterior border of the postero-superior offshoot.

5. Pathways in the form of a narrow band may be seen to extend between a part of the superior border adjoining the postero-inferior offshoot and the anterior limb of the bifurcation of the postero-inferior offshoot.

6. Between the upper extremity of the sulcus lunatus and the posterior branch of the bifurcation of the postero-inferior offshoot.

7. Pathways across the sulcus lunatus; across the postero-superior offshoot; across the postero-inferior offshoot.

2. The Superior Segment of the Postcentral Fissure
(Figure 22)

Orientation.—This was continuous on the surface with the inferior segment, but was separated from it in the depth of the fissure by a prominent annectant already mentioned. It had two small anterior offshoots, which incised the postcentral convolution, and two larger offshoots posteriorly. Above, the fissure incised the superior border of the hemisphere and passed for a short distance on the mesial surface as the posterior boundary of the paracentral lobule. Three annectant buttresses of slight prominence marked its upper course, two being on its anterior wall and one on the posterior wall.

Pathways.—1. Between the antero-inferior offshoot and (a), the antero-superior offshoot; (b), the mesial extremity of the fissure.

2. Between a higher portion of the middle third of the posterior border of the fissure and the upper border of its postero-superior offshoot.

3. Between the lower portion of the middle third of the anterior border and (a), the inferior annectant which separates this fissure from the inferior postcentral; (b), the inferior border of the postero-inferior offshoot.

4. Pathways whose direction is transversely around the fissure throughout its length. In the situation of each of its several offshoots, these pathways extend between the particular offshoot and the opposite border of the fissure.

3. The Posterior Ascending Limb of the Sylvian Fissure
(Figure 22)

Orientation.—In the hemisphere studied, this fissure consisted of two portions: an inferior diagonal sulcus, directed from below and behind upwards and forwards; and a superior vertical sulcus, directed from below and in front upwards and somewhat backwards. Below, the vertical anastomosed with the diagonal portion; above, with the arch of the interparietal fissure; and behind, with the temporo-parietal fissure. It was separated from all three fissures by annectant convolutions.

Pathways of the diagonal portion.—1. Between the antero-superior extremity and (a), the lower third of the floor of the fissure; (b), a diagonal fissure placed on the parieto-temporal operculum (see Figure 34).

2. Between a diagonal fissure on the parieto-temporal operculum (Figure 34) and (a), the upper half of the posterior border; (b), the annectant which separates this fissure from the vertical portion.

3. A few pathways which extend more or less transversely across the fissure in its upper and lower extremities.

Pathways of the vertical portion.—1. Between an inferior offshoot and the superior border.

2. Between the superior annectant and the postero-inferior border.

3. Between the antero-superior border and (a), the posterior annectant; (b), the postero-inferior border; (c), the lower border of an anterior offshoot.

4. Between the opposite borders of the fissure at its highest part.

5. Between the inferior annectant and a small extent of the lowest part of the posterior border.

6. Between the inferior annectant and (a), a portion of the posterior border; (b), the posterior annectant.

Above these pathways and parallel to them, fiber bundles are shown on the map as extending between the middle third of the anterior border and the posterior annectant.

7. Between the entire lower border of the anterior offshoot and the anterior half of its upper border.

A few transverse pathways around the anterior offshoot near its middle.

4. The Temporo-Parietal Fissure (Figure 22)

Orientation.—From the point of view of the intercortical pathways only that portion of the fissure can be conveniently described under the heading of temporo-parietal which, in the hemisphere studied, was interposed between the interparietal arch (the antero-inferior offshoot of the parietal ramus of the sulcus occipitalis transversus) and the ascending (parietal) limb of the first temporal fissure. On the one hand it was separated from both these fissures by a prominent annectant convolution; on the other hand, its contained pathways were so thoroughly interlaced with those of an anterior tributary sulcus, that they cannot be separated in the enumeration of the intercortical systems. By way of this tributary sulcus the temporo-parietal fissure of the hemisphere under description communicated on the surface with the posterior ascending limb of the Sylvian fissure. The sulcus was, however, separated from the latter by an annectant convolution. Antero-superiorly the communicating sulcus terminated in the form of an offshoot. A smaller offshoot may be discerned on the map projecting from the posterior part of the main fissure, by the side of the superior annectant.

Pathways.—1. Between the inferior annectant and (a), the anterior annectant; (b), the anterior offshoot; (c), the posterior offshoot; (d), the posterior border of the fissure.

2. Between the posterior border of the fissure and the anterior offshoot of the communicating sulcus.

3. Between the superior annectant and the posterior part of the inferior border of the communicating sulcus.

4. Between the anterior annectant and the posterior part of the superior border of the communicating sulcus.

CHAPTER NINE

INTERCORTICAL SYSTEMS OF THE CENTRAL AREA

1. The Central Fissure (Figures 21 and 22)

The attention of the older morphologists was especially attracted to the more or less vertically placed central fissure and convolutions of the primate cerebrum, by reason of the interruption caused by these structures in the generally horizontal course of the fissures and convolutions in the rest of the cortical extent. So persistent had been the idea of a particular significance attaching to this fact, that long before Fritsch and Hitzig's discovery of a muscular response to the excitation of the precentral gyrus, and of Betz' finding of the giant pyramidal cells, the anatomical literature already contained special references to the central fissure and convolutions. An interesting point in connection with the literature on the subject was brought out in a lecture by Broca (41): it is an account of the singular literary accident which led Leuret, in 1839, to attach Rolando's name to the central fissure, notwithstanding the fact that Rolando himself merely mentioned it as having been described forty years before by Vicq-d'Azyr and later by Gall and Spurtzheim.

For a thorough description of the gross forms of the several types of the central fissure, anatomy is indebted to Eberstaller, to whose book (13) the reader is referred. According to Cunningham (42) the fissure develops but rarely as a single furrow. Most frequently a lower furrow appears first, soon followed by the appearance of an upper one, the coalescence of the two forming the central fissure of the adult. The site of union of the two furrows out of which it is formed is generally well marked in the adult, at or near the junction of the upper and middle thirds, either by an annectant convolution or by interlocking buttresses which meet above the floor and raise it to a variable height. The latter is an important physiological landmark (43). In 1,667 brains, Heschl (11) found the central fissure entirely bridged over by the annectant convolution in question in 6 hemispheres; the annectant rose to from one-third to five-sixths the depth of the sulcus in 67 hemispheres; and from one-sixth to one-third the depth in 75 hemispheres. Cunningham (42) found this annectant conspicuous in the five Negro brains which he examined.

> Not only is the condition more distinctly marked than in the European [says he], but in one hemisphere, taken from a young Timanee negress, the bridging gyrus is so strongly developed, that it all but reaches the surface. In the chimpanzee this deep annectant gyrus appears to be commonly present. The four hemispheres which I possess show it in a pronounced form in each case. The orang brain likewise gives evidence of the same condition, but not so distinctly.

Apparently it was this author's opinion that the prominence of the annectant convolution spoke for a correspondingly low grade of cerebrum. But Wagner's (44) finding of a complete interruption of the central fissure in Professor Fuchs' brain makes one doubt the validity of such an opinion. Eberstaller found a complete bridging of the fissure by a convolution in 2 out of 200 brains. Cunningham has never met with it. I have been so fortunate as to find a hemisphere in which the central fissure was bridged over almost completely, and the contained intercortical pathways of this fissure will be shown later.

In from one-fifth to one-third of human hemispheres, the lower end of the central fissure communicates with the horizontal fissure. According to Eberstaller, with whom all the anatomists whose work I have consulted agree on this point, such a union of the two fissures is accomplished indirectly, by way of a transverse sulcus the greater part of whose length generally lies concealed on the operculum. The direction of these small sulci on the superior operculum varies greatly and Retzius has applied to them what appears to be the better name of "subcentral."

Cunningham states that in the cases in which the central fissure joins the horizontal, the union of the two is generally complete. But Symington and Crymble (45), who have made a special study of this point, say that

> those fissures which communicated with the Sylvian fissure, showed an elevation of the floor near the lower end. . . . The deep annectant gyrus may then be looked upon as the buried lower extremity of the central fissure.

I have examined a number of hemispheres regarding this point, and found every gradation between a complete convolution separating the central from the subcentral sulci and an entirely smooth lower end of a central fissure which communicated without interruption with the horizontal.

Orientation.—The central fissure of the left hemisphere pictured in Figure 22 was a rather simple variation of a common type. It incised the upper border of the hemisphere, passing backwards for a short distance on the paracentral lobule on the mesial surface. It was separated below from the horizontal fissure by a thin convolution. The lower genu of the fissure was well marked, but the upper was a mere wavy line which could hardly be described as a knee. The walls of the fissure were relatively smooth, the usual interlocking buttresses being so flat that they hardly deserved the name. The annectant gyrus above the middle of the fissure was but faintly marked. At the junction of the inferior and middle thirds, the central fissure communicated with the inferior segment of the precentral fissure, by way of a sulcus, but was separated from it by a deep annectant. Above the latter, at successive levels from below upwards, this central fissure had one offshoot behind, incising, but not severing, the postcentral convolution, and two offshoots in front, which incised the superior segment of the precentral convolution.

The pathways which traverse the central fissure are of especial interest for a number of reasons. The precentral convolution gives origin to the cortico-spinal tract and is essentially efferent in its function; while the function of the postcentral convolution is essentially afferent. Excitation of definite points on the former convolution constantly elicits movement of correspondingly definite groups of muscles, while no muscular response and but an inconstant and rather vague subjective sensory response (46) results from a similar excitation of the latter. In his epoch-making work on the cytology of the cerebral cortex, Campbell (47) searched for pathways which might connect the two structurally different and, in the light of all reason, functionally interdependent, convolutions. He was unable to find them in the cortex, but he made the important suggestion that if such pathways exist, they must be searched for beneath the cortex. At different levels in the latter location such connecting pathways have been found and mapped in the present study. Their somewhat different courses in different types of the central fissure of the human cerebrum will first be described. The pathways which connect the central cortex with more distant parts of the cerebrum will be described in connection with the arcuate fasciculus.

Pathways of the central fissure (Figure 22).—1. Pathways directed diagonally around the fissure, from behind and below, upwards and forwards: Between points on the entire anterior border (precentral convolution), including the two anterior offshoots as well as the annectant which separates this fissure from the inferior precentral and points at lower levels on (a), the inferior extremity (superior operculum) of the fissure; (b), the entire length of the posterior border (postcentral convolution), including the posterior offshoot.

2. Pathways which extend in the form of a narrow band between a small part of the posterior border, immediately below the posterior offshoot and a small part of the anterior border, a short distance above the superior annectant.

3. Between the upper part of the middle third of the anterior border (precentral) and the antero-superior offshoot.

4. Between the floor of the upper fourth of the fissure and its mesial (paracentral) extremity. These pathways are apparently continuous with those which extend between a part of the middle third of the anterior border and the floor of the upper fourth of the fissure.

5. Between the lower portion of the middle third of the anterior border and the same border of the highest portion of the fissure immediately above the antero-superior offshoot.

6. Pathways between a short extent of the anterior border, below the anterior communicating sulcus, and a short extent of the posterior border above the posterior offshoot.

7. Vertical pathways across the posterior offshoot.

8. Pathways squarely or slightly diagonally around the fissure throughout its length, extending between the precentral and postcentral convolutions. In the situation of the anterior communicating sulcus these pathways extend by the sides of the anterior annectant and some beneath its base (not shown on the map), continuing their course in the bed of the inferior precentral fissure as far as its anterior border (the middle frontal convolution). In the lower fourth of the fissure the course of these fiber bundles is at a somewhat different angle, being from behind and above, forwards and downwards, so that they cross the transverse bundles above them for a short space in the posterior wall of the fissure.

9. Another set of transversely directed bundles is intermixed with the foregoing in the lowest portion of the fissure. It is in the form of a narrow band which is somewhat curved with the convexity upwards.

The Central Fissure of the Right Hemisphere of the Cerebrum, from The Left Hemisphere of Which the Large Maps Were Made
(Figures 23, 24 and 25)

The central fissure of this hemisphere extended above as far as the superior margin; it barely incised the opercular margin below. Its two genua were well marked. The buttresses on its walls were shallow. Its floor was slightly elevated in two or three places but nowhere into a prominence deserving of the name of annectant convolution.

The fissure was bifurcated above, the posterior branch of the bifurcation being the continuation of the fissure, the anterior one being an offshoot which incised the precentral convolution. Another offshoot which incised the latter convolution was at the junction of the superior and middle thirds of the fissure. By way of two sulci which severed the precentral convolution, this fissure communicated with the inferior precentral. Two offshoots at about the middle third of its posterior border incised the postcentral convolution.

Pathways.—1. In the long axis of the fissure between its two extremities.

2. Diagonal pathways around the fissure between points on its posterior border and points at higher levels on its anterior border, including the offshoot and the communicating sulci.

3. Diagonal pathways around the fissure between points on its anterior border and points at higher levels on its posterior border.

Note.—Three or four spaces may be observed in the composite picture (Figure 25) which are devoid of diagonal lines. From the experience gathered from a number of other fissures of the relation of these pathways to annectant convolutions and buttresses, it is safe to conclude that these spaces correspond to the shallow elevations on the floor and walls of this fissure.

2. Rare Types of the Central Fissure

In contrast to the relatively simple types of the central fissure just described, there are others of a more complicated and, for the present, largely inexplicable character.

The pathways of the two central fissures pictured in **Figures 26, 27, 28, 29 and 30** are respectively from the left and right hemispheres of one cerebrum.

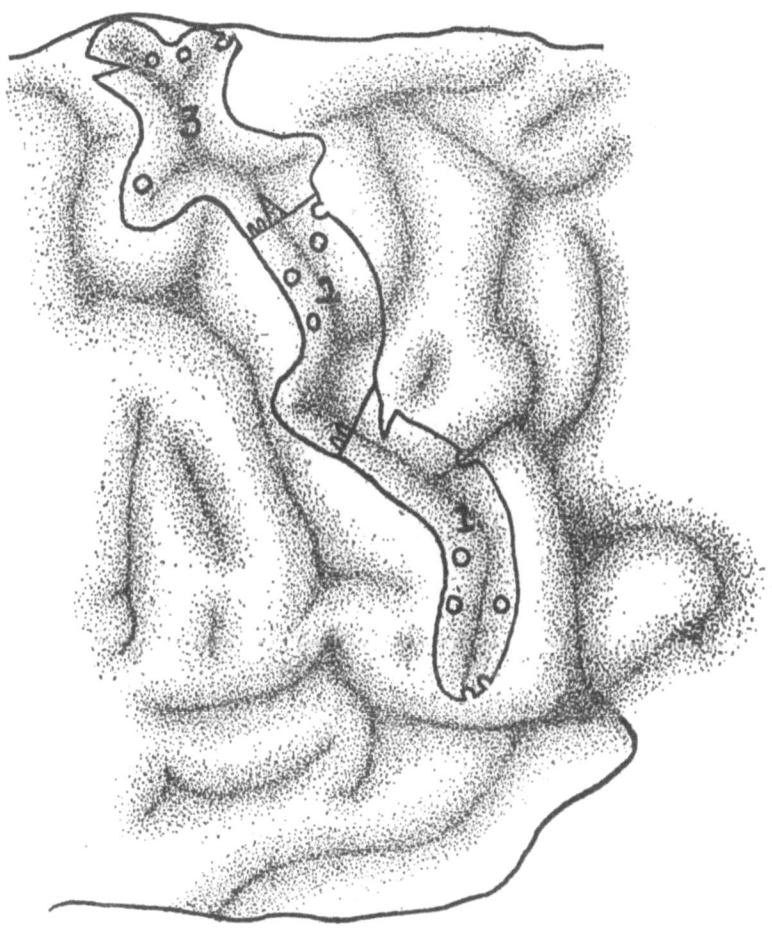

FIG. 23. THE CENTRAL FISSURE OF THE RIGHT HEMISPHERE OF THE CEREBRUM (FROM THE LEFT HEMISPHERE OF WHICH THE LARGE MAPS WERE PREPARED)
Drawing from a plaster-of-Paris cast

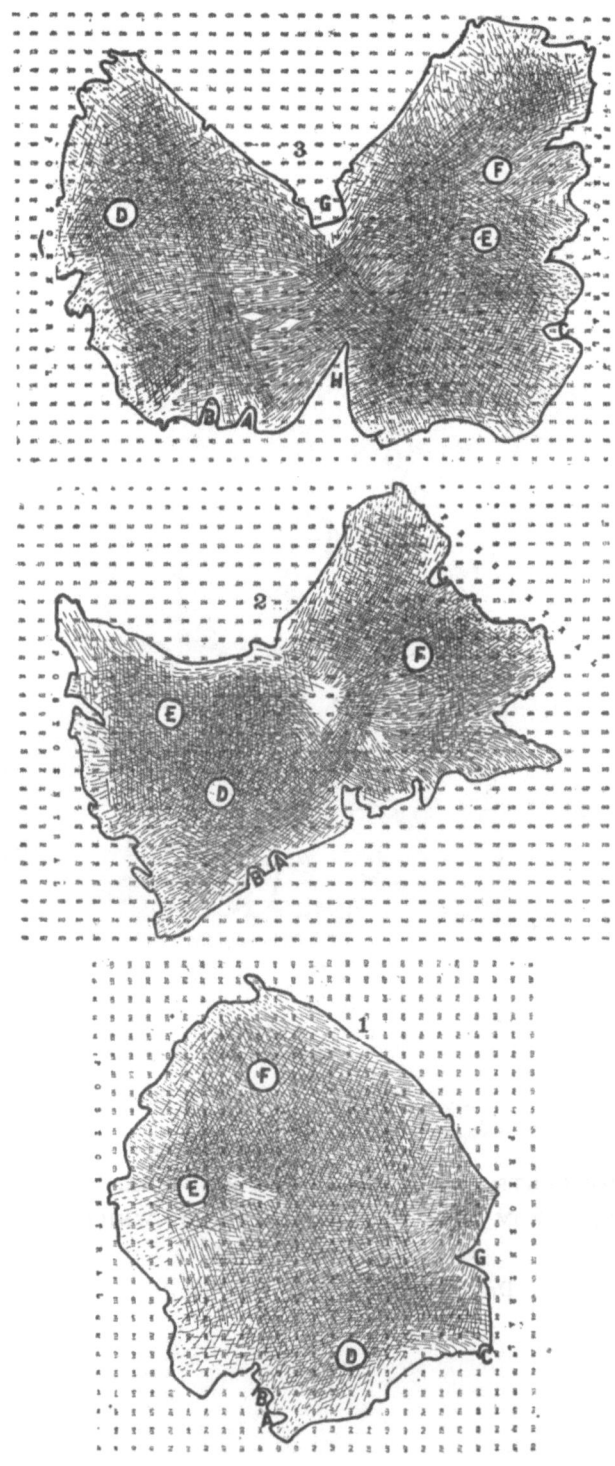

FIG. 24. DETAILED DRAWINGS OF SECTIONS (MAGS. 50 AND 150) OF THREE PARTS OF THE SUBCORTICAL LAMINA OF THE RIGHT CENTRAL FISSURE SHOWN IN FIGURE 23, ILLUSTRATING THE METHOD OF RECONSTRUCTION (SEE CHAPTER II, 6)

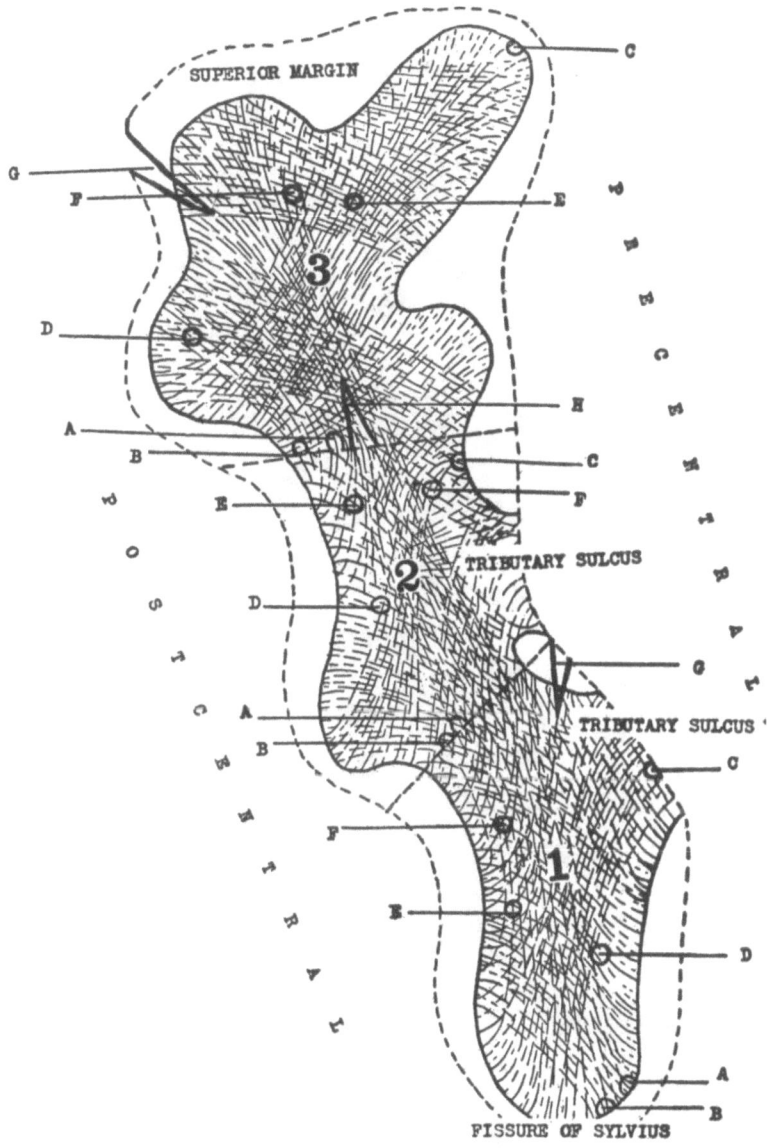

FIG. 25. A SCHEMA OF THE INTERCORTICAL PATHWAYS OF THE CENTRAL FISSURE OF THE RIGHT HEMISPHERE OF THE CEREBRUM (FROM THE LEFT HEMISPHERE OF WHICH THE LARGE MAPS WERE PREPARED) RECONSTRUCTED FROM THE DETAILED DRAWINGS IN FIGURE 24

In the reconstruction the worker was guided entirely by the relation of the lines to the original recorded markings.

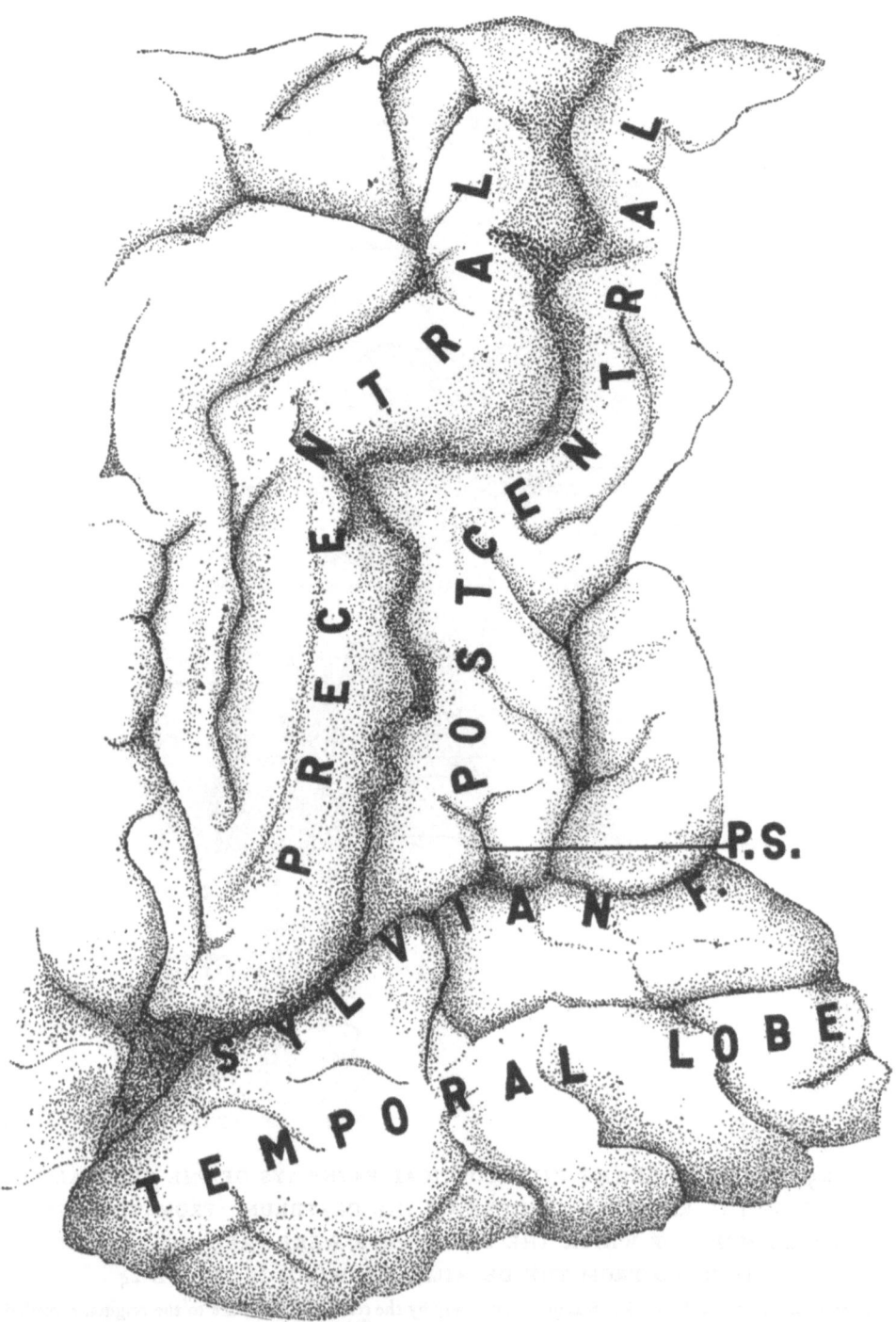

FIG. 26. A RARE TYPE OF THE LEFT CENTRAL FISSURE

Drawing from a plaster-of-Paris cast. In the preparation of the mold for the cast, the central fissure was partly opened. (P. S., posterior subcentral fissure.)

The fissure of the left hemisphere (Figure 26) was remarkably long, extending from the paracentral lobule on the mesial surface above to the horizontal fissure below. The pronounced curvature of both its genua contributed further to its great length. Its interlocking buttresses were numerous and prominent. But there was not a sign of an annectant gyrus either at the junction of the upper and middle thirds, or below, near the horizontal fissure. In the hemisphere in question there was a well-marked posterior subcentral but not a trace of an anterior subcentral sulcus, which must have been thoroughly incorporated with the lower end of the central fissure. The latter fact, together with the utter absence of any annectant gyri in the fissure, speaks for an unusually high degree of fetal development. Attention is directed to these facts because of the remarkable contrast presented by the central fissure of the right hemisphere of the same cerebrum. To anatomists who are of the opinion that prominent annectants are significant of a low grade of cerebrum, the facts presented must be of special interest.

The central fissure of the right hemisphere (Figure 29) was remarkably straight and short. It reached neither the cerebral border above, nor the horizontal fissure below. It was almost completely separated into two parts at the junction of the upper and middle thirds by a convolution, only a small part of which lay beneath the surface. The form of this convolution was that of the letter V. The right limb of the V, which was much the longer, was on the surface. The left was buried to a depth of about 5 millimeters. The upper segment of the fissure communicated in front with the superior precentral fissure; behind it sent an offshoot, which incised deeply, but did not sever, the postcentral gyrus. The lower segment had, about its middle, an anterior and a posterior offshoot, not quite opposite each other, and three interlocking buttresses.

Pathways of the left central fissure (Figure 27).—1. Pathways in the long axis of the fissure.

2. Diagonally across the upper half of the fissure between points on the postcentral border and points at higher levels on the precentral border. A narrow band of such pathways may be seen on the map, in the lower third of the fissure.

3. A narrow band squarely around the fissure at the junction of its middle and upper thirds.

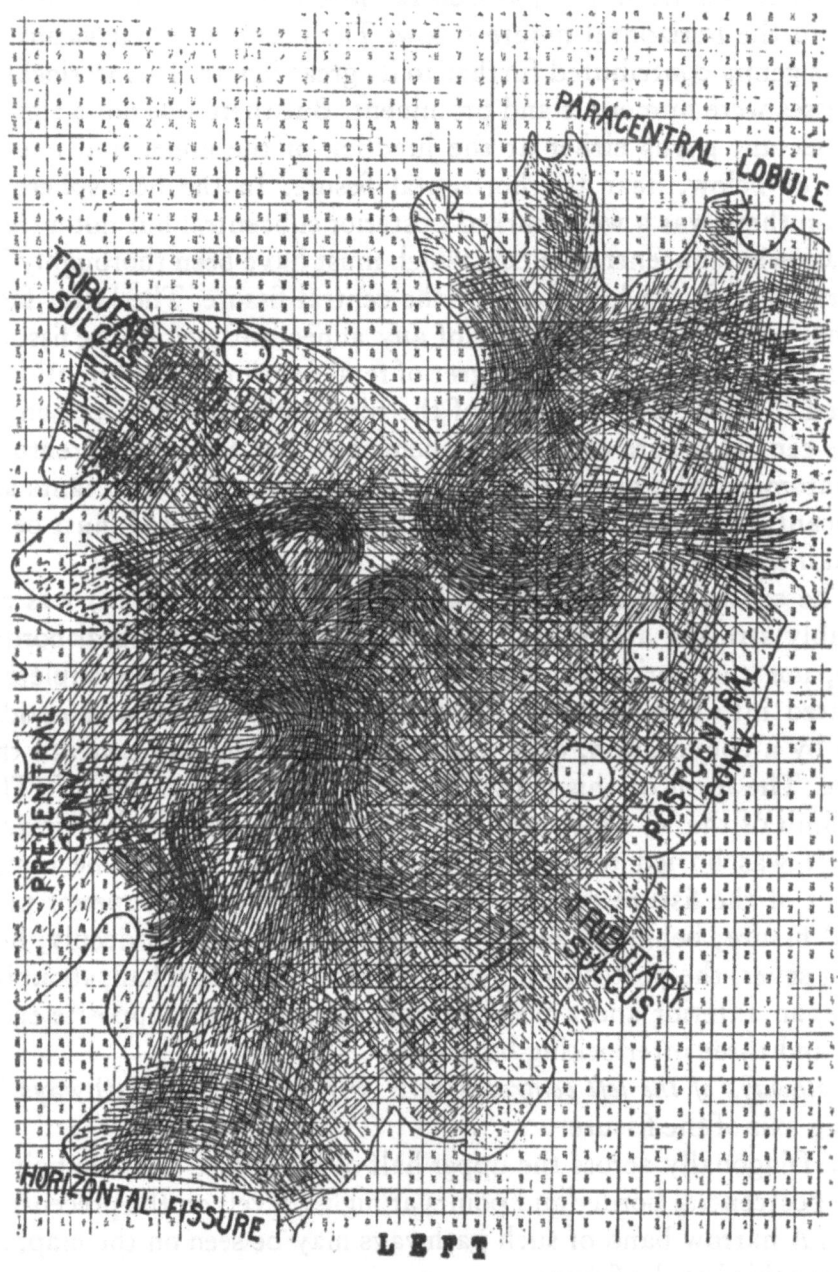

FIG. 27. THE INTERCORTICAL PATHWAYS OF THE LEFT CENTRAL FISSURE SHOWN IN FIGURE 26

FIG. 28. A PHOTOMICROGRAPH (X 130) OF THE INTERCORTICAL HAIRPIN PATHWAYS FROM THE SUBCORTICAL LAMINA OF THE LEFT CENTRAL FISSURE SHOWN IN FIGURE 27

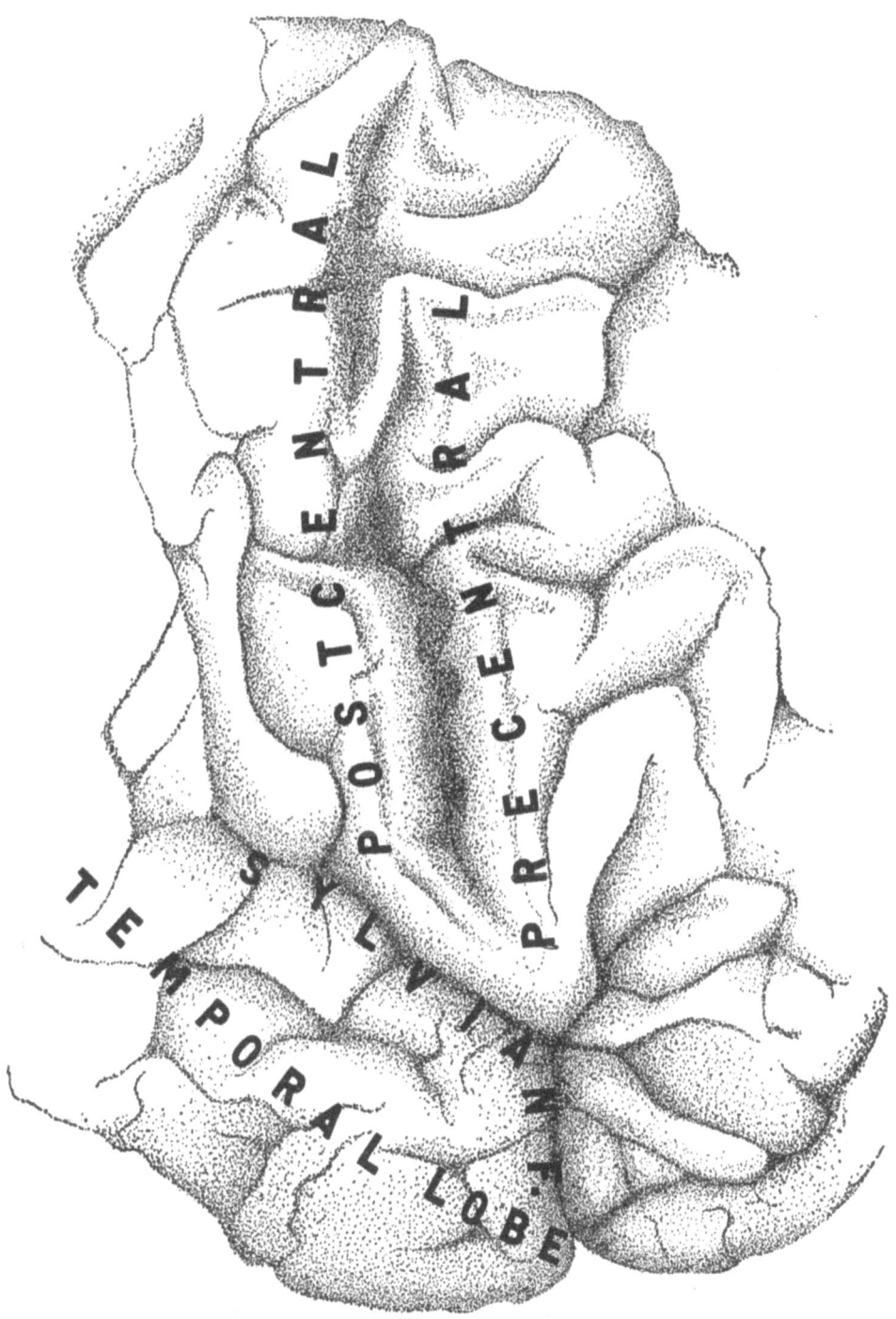

FIG. 29. A RARE TYPE OF THE RIGHT CENTRAL FISSURE

Drawing from a plaster-of-Paris cast. In the preparation of the mold for the cast, the central fissure was partly opened.

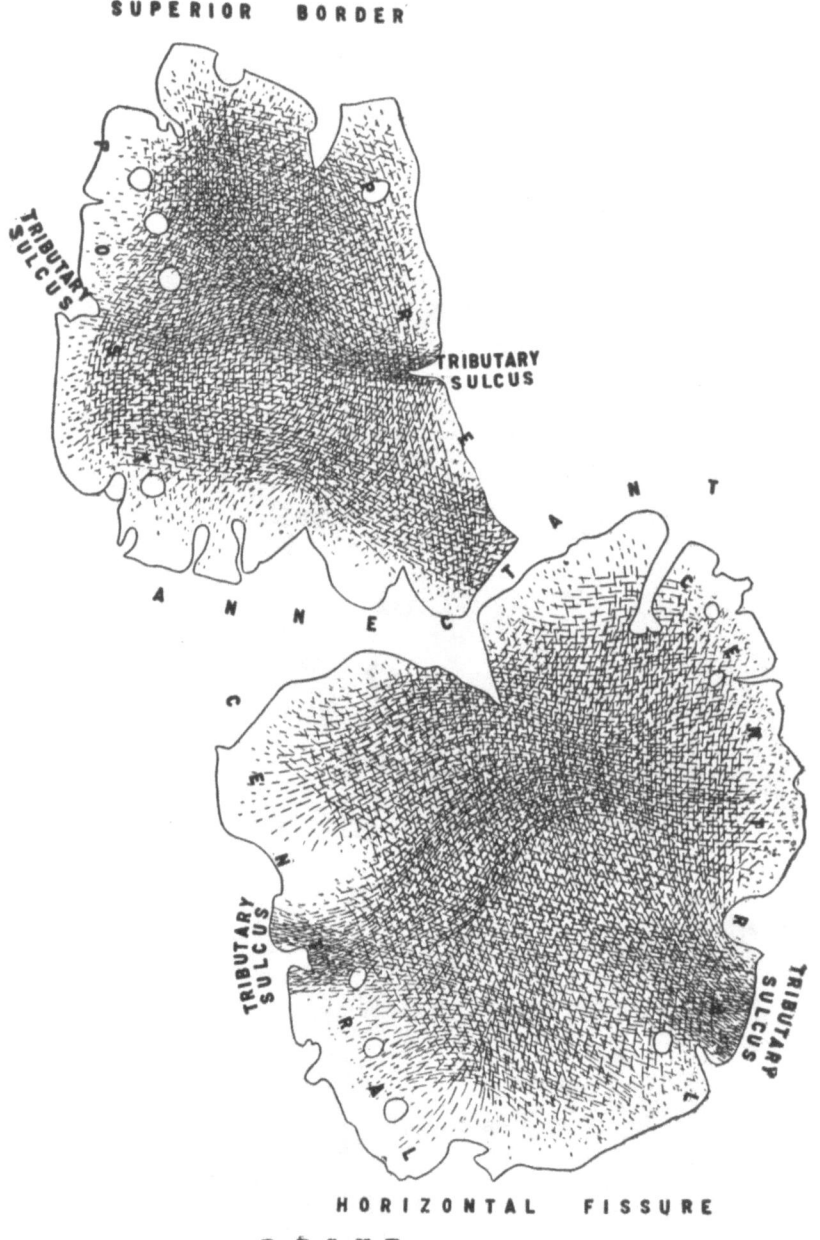

FIG. 30. THE INTERCORTICAL PATHWAYS OF THE FISSURE SHOWN IN FIGURE 29

4. A narrow band between the posterior wall of the paracentral portion of the fissure and the floor of the middle of the fissure.

5. Seven or eight strands, each bent more or less in the form of a hairpin (Figures 27 and 28) in different parts of the fissure.

6. A narrow, slightly curved band, with the convexity upwards, in the lowest portion of the fissure.

Pathways of the right central fissure (Figure 30).—In the superior segment:

1. Between the superior extremity and the annectant convolution.

2. Diagonal pathways between points on the postcentral border and points at higher levels on the precentral border. Parallel with these, pathways extend between the annectant and the lower half of the precentral border.

3. Almost squarely around the fissure between the precentral and postcentral borders.

4. Between the tributary sulci, around the fissure, in curves with an upper convexity. The highest contingents of these pathways extend between the anterior tributary and the upper half of the posterior border.

In the inferior segment:

1. Between the anterior tributary sulcus and the anterior half of the annectant.

2. Pathways parallel to the above between the inferior extremity of the fissure and the posterior half of the annectant.

3. Between the inferior extremity and upper half of the anterior border. Parallel with these there are pathways between the posterior border and the entire annectant.

4. Between the posterior tributary sulcus and the anterior wall of the inferior extremity.

5. Between the anterior tributary and the posterior half of the annectant.

6. A narrow band between the posterior tributary and a small extent of the middle of the anterior border. These pathways are curved with the convexity upwards. Parallel to these pathways, others extend around the upper half of the fissure in gradually diminishing curves.

CHAPTER TEN

INTERCORTICAL SYSTEMS OF THE FRONTAL AND PREFRONTAL AREAS

1. The Topography of the Frontal Lobe

Of all the areas of the cerebrum, the topography of the frontal and prefrontal areas is most subject to variation. In 57 hemispheres, for example, which Cunningham examined, there were 14 different varieties of the sulcus frontalis secundus. These resulted from different combinations of the parts of that sulcus and the differences in the disposition and the degree of prominence of the annectant convolutions. In 69 hemispheres the same investigator found 27 different varieties of Eberstaller's sulcus frontalis medius—varieties which differed from each other with respect to their direction and their different combinations with other fissures. Since a marked variation of any one fissure with respect to its length, direction and anastomoses must likewise affect the other fissures in the area which it traverses, it will be readily appreciated how the topography of an entire lobe may, even in the eyes of an expert, assume an unfamiliar appearance, as the result of some irregularity of a single fissure.

The apparently abnormal topography of the frontal lobe seen in Figures 21 and 31 is due mainly to the sulci which traverse the area between the superior and inferior frontal fissures. It is not an easy matter to decide which of the fissures of that area represents Eberstaller's sulcus frontalis medius. It is possibly represented by more than one sulcus on the hemisphere pictured, being, in morphological language, "divided parts of a single fissure."

2. The Superior Precentral and the Superior Frontal Fissures
(Figure 31)

Orientation.—The two fissures together contained five annectants which prove of assistance in the orientation of their pathways: From the front backwards, the first annectant separated the anterior bifurcated extremity of the fissure. Between the first and second annectants the superior frontal fissure communicated below with a long diagonal fissure which probably represented Eberstaller's middle frontal. Between the second and third annectants the fissure had one offshoot above and one below. The fourth annectant, placed horizontally, marked the partial separation of a superior trifoliate offshoot. Between the fourth and fifth annectants, the superior frontal communicated below with the superior precentral. The fifth annectant marked off the posterior bifurcated extremity.

It will be observed that no annectant intervened between the junction of the frontal and precentral. On the surface the anastomosis between the two was complete and, as may be seen on the map (Figure 31), the intercortical pathways extend between the two fissures without any interruption. A review of the literature on the subject discloses the fact that the anastomosis of these two fissures is one of the most stable features of the surface topography of the human cerebrum. Eberstaller indeed considered the anastomosis of these two fissures to exist uniformly without any exception. Cunningham, however, has observed rare instances in which there existed a separation by a deep and even by a superficial gyrus.

Pathways.—1. In the space between the anterior extremity and the second annectant:

Between the superior border of the superior frontal and the anterior border of the middle frontal.

2. Throughout the length of the superior frontal, in part interrupted by the several annectants in its course; between the inferior branch of its anterior bifurcation and the middle third of the posterior border of the superior precentral.

In the space between the second and third annectants:

3. Between the inferior border of the fissure posterior to the second annectant, including the inferior offshoot and the superior border of the fissure between the second and third annectants.

4. Between the anterior border of the superior offshoot and (a), the third annectant; (b), the inferior border of the fissure anterior to the third annectant; (c), the posterior border of the inferior offshoot.

In the trifoliate tributary:

5. Between the posterior border of the central leaf and (a), the anterior border of the central leaf with a downward and forward sweep; (b), the anterior border of the central leaf, in a nearly vertical direction; (c), the inferior border of the anterior leaf; (d), the third annectant.

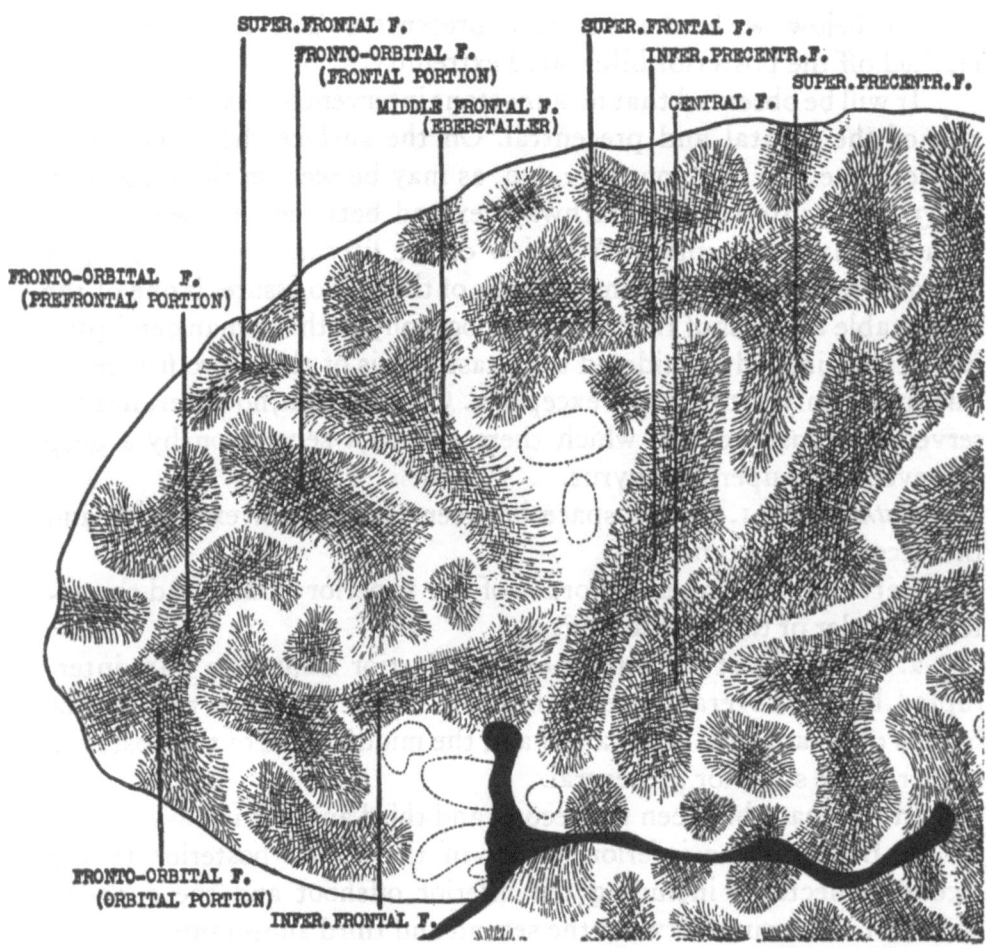

FIG. 31. A COMPOSITE SCHEMA OF THE INTERCORTICAL SYSTEMS OF THE LEFT FRONTAL AND PREFRONTAL AREAS OF A HUMAN CEREBRUM

6. Between the fourth annectant and each of the three leaves of the trifolium.

7. Between the opposite borders of the posterior leaf.

In the portion of the fissure posterior to the third annectant, including the superior precentral and ignoring the interruption by the fifth annectant:

8. Between the inferior border of the fissure and (a), the third annectant; (b), the superior border of the fissure in the space between the third annectant and the trifoliate tributary.

9. Between the fourth annectant and (a), the inferior border of the fissure in front of the inferior precentral; (b), the inferior border of the fissure, behind the inferior precentral; (c), the lower extremity of the superior precentral.

10. Between the posterior bifurcated extremity of the fissure and (a), the anterior border of the superior precentral; (b), the anteroinferior border of the junction between the superior precentral and the superior frontal.

11. Vertical pathways in the posterior bifurcated termination of the fissure.

12. Between the superior border, immediately behind the trifoliate tributary and the inferior branch of the posterior terminal bifurcation.

3. The Inferior Precentral and the Inferior Frontal Fissures (Figure 31)

Orientation.—The upper part of the inferior precentral fissure was placed in this hemisphere in front of the superior precentral. It communicated behind with the central, in front and below, with the inferior frontal, but was partly separated from these fissures by annectant convolutions.

Pathways of the inferior precentral.—1. Between the posterior annectant which separates this fissure from the central and (a), the superior extremity of the fissure; (b), the upper half of its anterior border; (c), the upper half of its posterior border.

2. Between the middle third of the anterior border and the inferior extremity.

3. Slightly diagonal pathways around the fissure from points on its anterior border to somewhat higher points on its posterior border, including the posterior annectant.

4. In the lower two-thirds of the fissure pathways extend between points on the anterior border and points at lower levels on the posterior border, including the posterior annectant. The continuation of these pathways in the bed of the central fissure may be plainly seen on the map.

Orientation.—It will be seen on the map that the inferior frontal fissure was connected on the surface behind with the inferior precentral; behind and above, with Eberstaller's middle frontal; and in front, with a diagonal or radiate sulcus. It was partly separated from all these fissures by annectants. The course of most of the pathways contained in the inferior frontal and in the diagonal sulcus coincided to such an extent, that the two fissures may be considered under one heading.

Pathways.—1. Between the anterior annectant and (a), the posterior annectant: (b), the postero-superior annectant.

2. Between the inferior border of the inferior frontal and (a), the posterior annectant; (b), the posterior border below the posterior annectant.

3. Between the anterior half of the inferior border of the inferior frontal and (a), the inferior branch of an anterior bifurcation of the diagonal sulcus; (b), the anterior half of the inferior border of the diagonal sulcus.

4. Between the superior border of the inferior frontal fissure and (a), the lower half of the posterior border; (b), the inferior border of the inferior frontal fissure; (c), the anterior ascending limb of the fissure of Sylvius.

5. Between the middle two-fourths of the lower border of the diagonal sulcus and (a), a small anterior portion of the upper border of the inferior frontal; (b), the superior border of its junction with the inferior frontal.

6. Curved pathways between the anterior annectant, which separates the inferior frontal from the diagonal sulcus and the postero-inferior extremity of the diagonal sulcus.

7. Pathways diagonally around the postero-inferior extremity of the diagonal sulcus.

8. Between the superior border of the anterior bifurcation and the postero-inferior extremity of the diagonal sulcus.

4. The Middle Frontal Fissure (Eberstaller)

In the hemisphere described, this fissure pursued an unusual course, being directed diagonally from behind and below, upwards and forwards. Above it anastomosed with the superior frontal; below, with the inferior frontal fissure, from which it was separated by an annectant convolution.

Those of its pathways which extend between it and the superior frontal have been enumerated in connection with the latter fissure. The only other pathways mapped in the middle frontal are those which extend around it from border to border. Below, these pathways may be seen to extend between the anterior border and the annectant which separates the fissure from the inferior frontal.

5. Fronto-orbital Fissures (Figures 20 and 31)

Orientation.—The inconstant topography of the prefrontal region of the cerebrum frequently makes it impossible to adhere to a conventional nomenclature of the fissures which traverse it. Outside of the fronto-marginal and four other small furrows, the area in question was traversed in this hemisphere by a large and rather complicated sulcus, which, for convenience in enumerating its contained intercortical pathways, may be divided into three parts. The first, or superior, portion of the fissure was located in the frontal region, in front of and parallel to the middle frontal fissure described; its direction, therefore, was from behind and below upwards and forwards. This fissure was joined, at about its middle point, at a right angle, by a fissure whose downward and forward course lay in the prefrontal area. It bifurcated below, near the orbital margin. Both branches of the bifurcation passed on to the orbital area; one traversing that area in front, from the lateral margin to within a short distance of the gyrus rectus; the other, laterally, parallel to the lateral orbital margin to within a short distance of the temporo-orbital recess.

For the purpose in hand the three portions of this fissure may therefore be named as the frontal, the prefrontal and the fronto-orbital proper. The prefrontal was separated from the other two portions of the fissure by annectant convolutions.

Pathways.—Three sets of pathways may be seen on the map (Figure 31) in the frontal (posterior) portion of this fissure.

1. In the long axis of the fissure.

2. Squarely around, between the annectant which separates it from the prefrontal portion of the fissure and a part of the superior border opposite this annectant.

3. More or less diagonally directed pathways around the fissure. In front, the direction of these pathways, as they appear on the map, is almost horizontal; behind, almost vertical.

Three sets of pathways may be seen on the map in the prefrontal portion of the fissure:

1. Between the inferior annectant and (a), the superior annectant; (b), the upper third of the anterior border; (c), the lower third of the posterior border.

2. Between the superior annectant and the postero-inferior border.

3. Around the fissure between its anterior and posterior borders.

The fronto-orbital portion of the fissure proper, and each of the two branches (Figures 20 and 31) into which it is bifurcated, may be seen to contain two sets of pathways; one set traverses the fissure in its long axis; the other extends more or less diagonally around it.

CHAPTER ELEVEN

DEEP INTERCORTICAL SYSTEMS OF THE LATERAL SURFACE OF THE CEREBRUM

1. The Arcuate Fasciculus and the Intercortical Systems of the Insular Region

Studies of the arcuate fasciculus date back to Burdach, who made the general observation that this fiber system served to interconnect the convolutions of the lateral surface of the cerebrum. Meynert (48) appears partly to have confused this fiber sheet with the deep intercortical sheet of the insula and its opercula. He speaks, however, of the arcuate fasciculus as consisting of two layers, a superficial one, located in the opercula of the Sylvian fissure, and a deep one, which extends from the temporal to the parietal region. Sachs (18) considered this system as part of his general stratum profundum convexitatis. Schnopfhagen (21) was of the opinion that the system in question consisted of fibers of the corpus callosum, which extended between the posterior region of one hemisphere and the anterior region of the other. The facts on which he based such a conclusion are not clearly given. Dejerine (19) concluded that the fasciculus in question consisted of short fibers which extended between neighboring convolutions. His schema of the arcuate fasciculus, which is in contradiction to the text, has been the subject of criticism by authors who otherwise agreed with his opinion regarding this fiber system. Dejerine and Thomas (49) reported a case of softening of the lower parietal region as far as the postcentral convolution. They were able to trace degeneration of the arcuate fasciculus in the first and second frontal convolutions. According to the authors, however, the tract of degeneration was indistinct as it approached the frontal gyri. Monakow (50) speaks of fiber bundles which extend between the first temporal and the foot of the third frontal gyrus. Anton and Zingerle's (25) observations agree with those of Dejerine's. Probst (20), devoted a considerable amount of study to this fiber system. He employed the Marchi method in experimental animals, as well as the Weigert-Pal method of massive degeneration. He arrived at the conclusion that the system of the arcuate fasciculus consists of relatively short fibers.

The arcuate fasciculus is frequently confused with the deep fiber sheet of the insula, and its opercula. The two fiber systems are entirely distinct. The deep insular fiber sheet has the form of a wide sac, whose opening corresponds to the fissure of Sylvius and whose sides and bottom correspond to the Sylvian opercula and to the sagittal plane of the island of Reil. The several directions of its constituent fibers may be studied in Figure 32, in which the sac, ripped open on its sides, is shown spread out on a flat plane. Its fibers, like the fibers of all "long" intercortical systems, are incorporated at their beginnings and ends in the subcortical laminae of the fissures of the region which they traverse. The capsule in question is therefore thoroughly adherent to the subcortical laminae of the fissures of the insula and the opercula (Figures 33 and 34) which, together with the layer of the cortex, line its inner surface.

The arcuate fasciculus, on the other hand, viewed sagittally, has the form of a wide horseshoe in whose opening lies the island of Reil, and whose upper convexity is nearly parallel to and about three centimeters below the superior margin of the cerebrum. Laterally it rests against the subcortical laminae of the fissures of the lateral surface of the cerebrum, in the space between the occipital and midfrontal areas. The mesial aspect of this horseshoe is smooth and concave; the lateral, convex and uneven, since it conforms to a certain extent to the furrowed plane of the lateral surface of the cerebrum.

Attempts to map the course of the fibers and fiber bundles of the arcuate fasciculus have been made in two ways:

1. The arcuate sheet was exposed on its mesial aspect by manual dissection of exploded brains (Figure 35). Curved slices were then cut through the entire thickness of the lateral cerebral shell in the apparent course of the arcuate bundles, that is, in the form of a wide horseshoe. These were then flattened on glass from above downwards; but as the slices had not only an inferior but a mesial concavity as well, the blocks, and the microscopic sections into which the blocks were made, remained mesially concave.

A detailed drawing of one of these sections is shown in Figure 36, from which the following facts may be gathered:

(a) The thickest part of the arcuate band is in the temporo-parie-

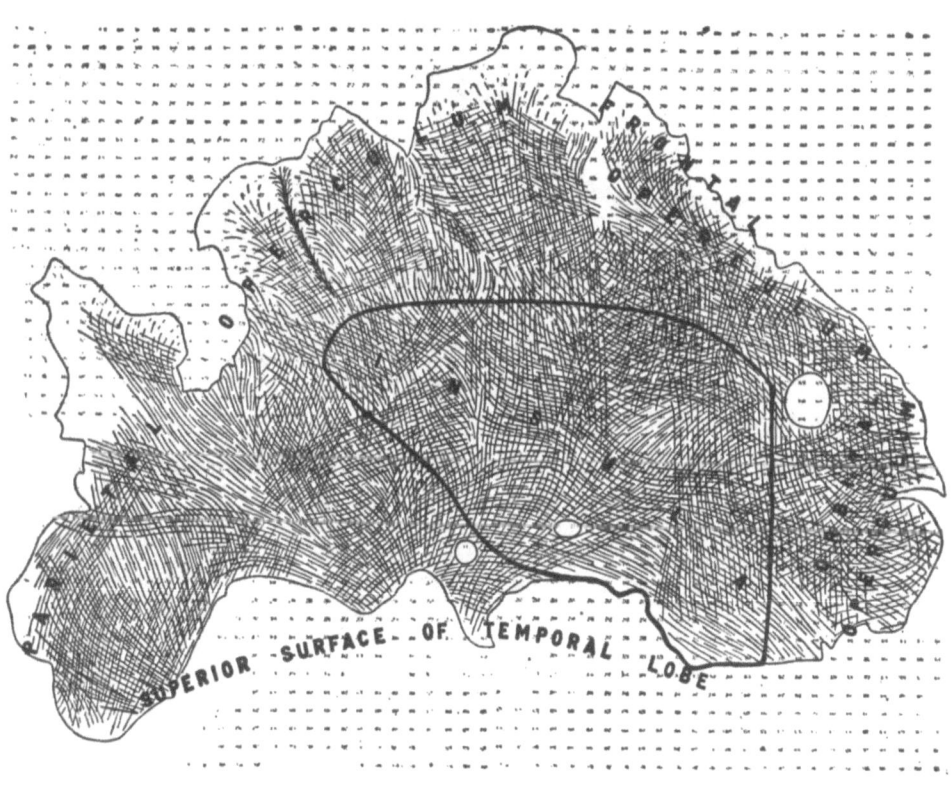

FIG. 32. INTERCORTICAL SYSTEMS OF THE DEEP FIBER SHEET LATERAL TO THE INTERCORTICAL SYSTEMS OF THE FISSURES OF THE INSULAR REGION

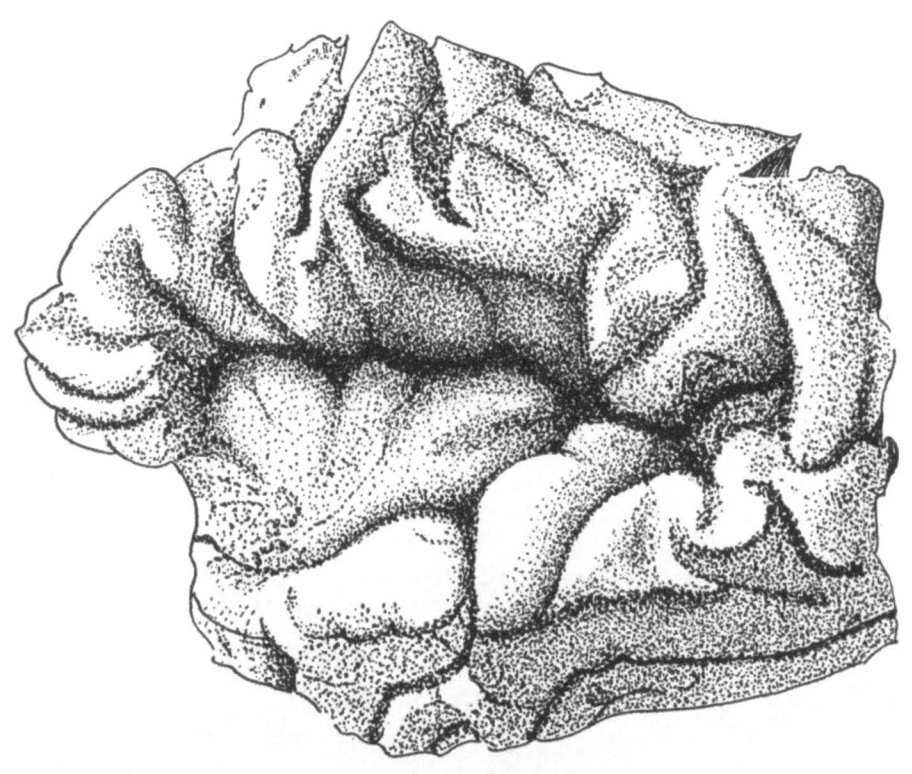

FIG. 33 A. THE INSULA AND THE OPERCULA OF THE LEFT HEMISPHERE STUDIED
Drawing from a plaster-of-Paris cast

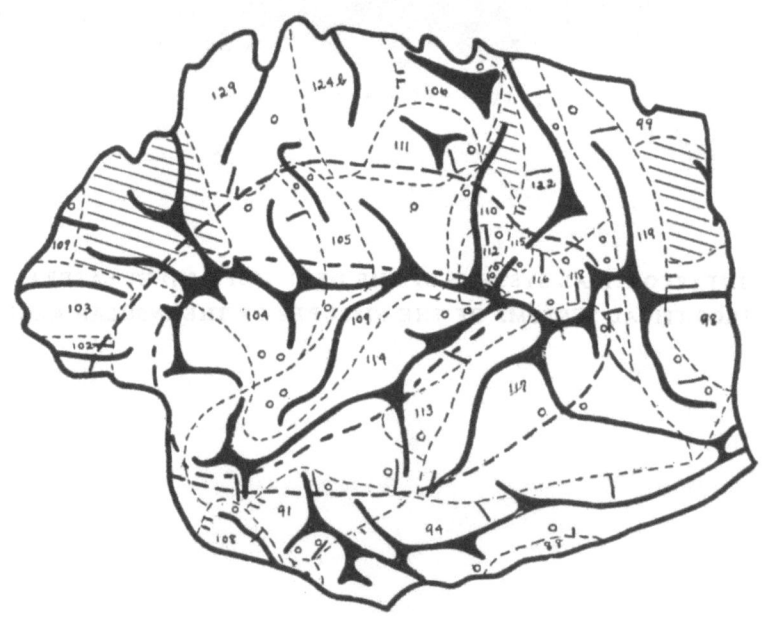

FIG. 33 B. A LINE DRAWING OF THE INSULA AND THE OPERCULA SHOWN IN FIGURE 33 A

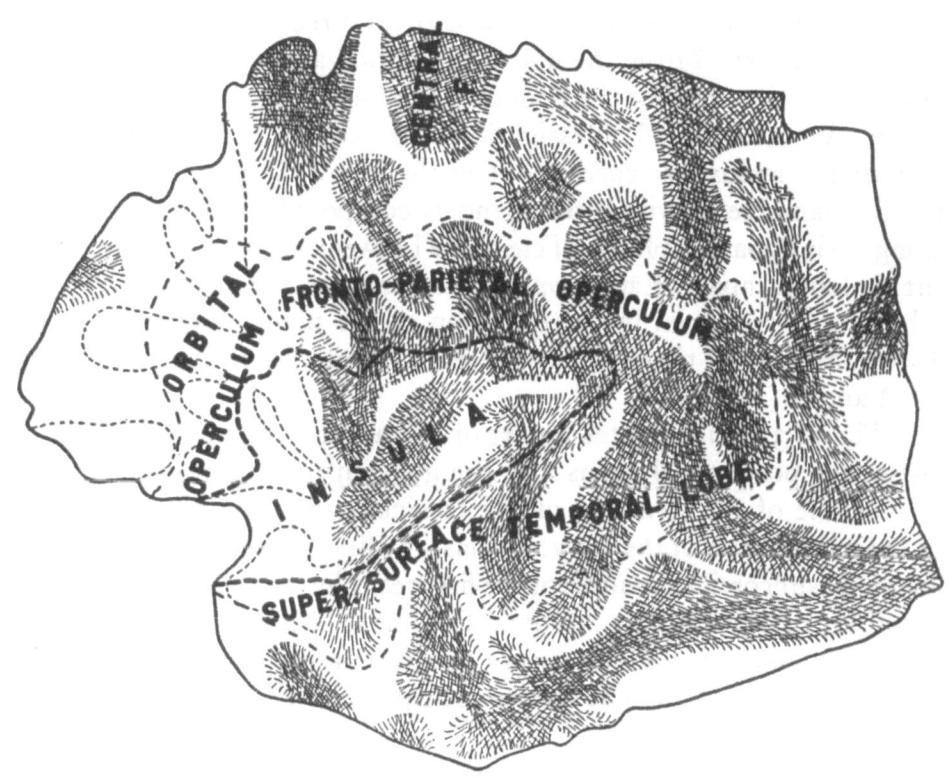

FIG. 34. A COMPOSITE SCHEMA OF THE INTERCORTICAL SYSTEMS UNDERLYING THE CORTEX OF THE FISSURES OF THE INSULA AND THE OPERCULA SHOWN IN FIGURES 33 A AND 33 B

tal region; the band thins out in the anterior parietal region and disappears from view in the precentral or posterior frontal.

(b) The most lateral fiber bundles are short and cannot be distinguished from the intercortical elements underlying the fissures; the more mesially placed bundles are increasingly longer.

(c) No fiber bundles may be seen to bend at a right angle from a sagittal to a lateral course, so as to aim for the crest of a convolution. Where fiber bundles deviate from a sagittal direction, the angle of the deviation is very obtuse, the line of projection aiming not for the crest of a convolution, but for the floor or wall of a fissure.

2. A more detailed investigation of the course and relations of the arcuate fasciculus was carried out in the following way. A number of the fine laminae of which this band is composed were dissected out along their apparent planes of cleavage between convolutions at different distances apart. Each lamina was flattened on glass, made into a block and sectioned in the direction of the plane. Examples of reduced drawings of the microscopic sections may be studied in Figures 37, A and B.

Figure 37 A illustrates a part of the arcuate band extending between the postcentral gyrus and the gyri supramarginalis and angularis. None of the elements of the fasciculus may be seen to extend between the crests of the convolutions, the line of their projection aiming at the subcortical laminae of the fissures. The more mesially situated bundles are increasingly longer. The most laterally situated bundles which underlie a given fissure do not enter (or emerge from) the subcortical lamina of that fissure, but proceed to join the lamina of a neighboring fissure.

Figure 37 B illustrates a procession of the band between the posterior portion of the inferior frontal region, two parieto-temporal gyri and the posterior portion of the second temporal gyrus. The fiber bundles in the space A–B are very much twisted and cannot be traced in continuity for any considerable length. At situation B a number of the bundles enter the subcortical lamina of the fissure. In the space B–C the laterally placed bundles aim for the subcortical laminae of the two fissures on each side of that which these bundles underlie. The more mesially placed bundles are longer and extend between fissures at greater distances apart.

In connection with the arcuate fasciculus a number of authors

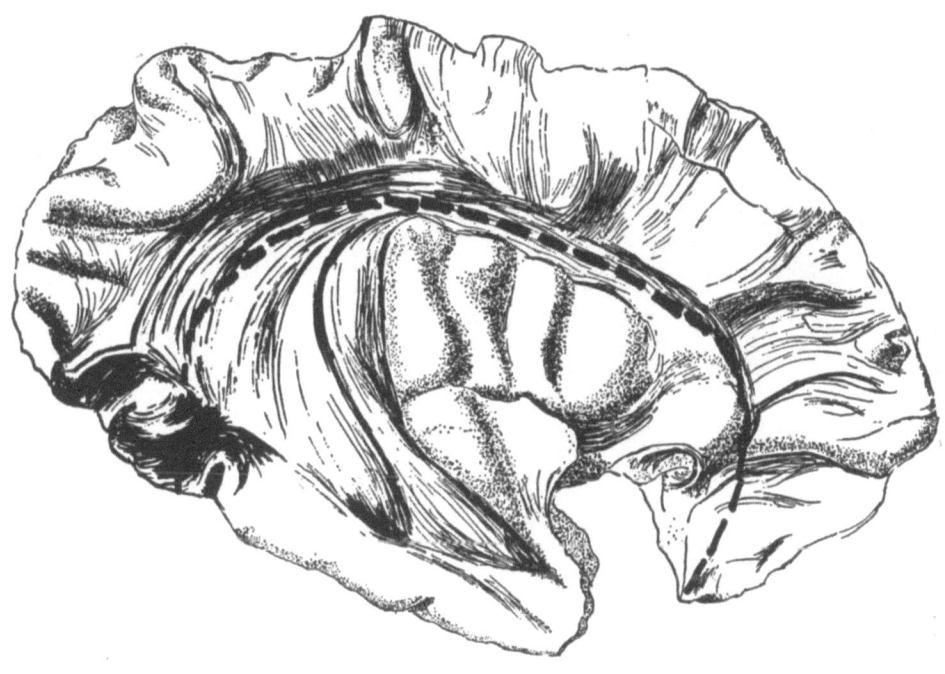

FIG. 35. A DISSECTION OF THE ARCUATE INTERCORTICAL SYSTEMS

The slices, from which the sections (Figure 36) were subsequently made, had been carried along the apparent course of the arcuate band indicated by the interrupted line. Drawing from a plaster-of-Paris cast.

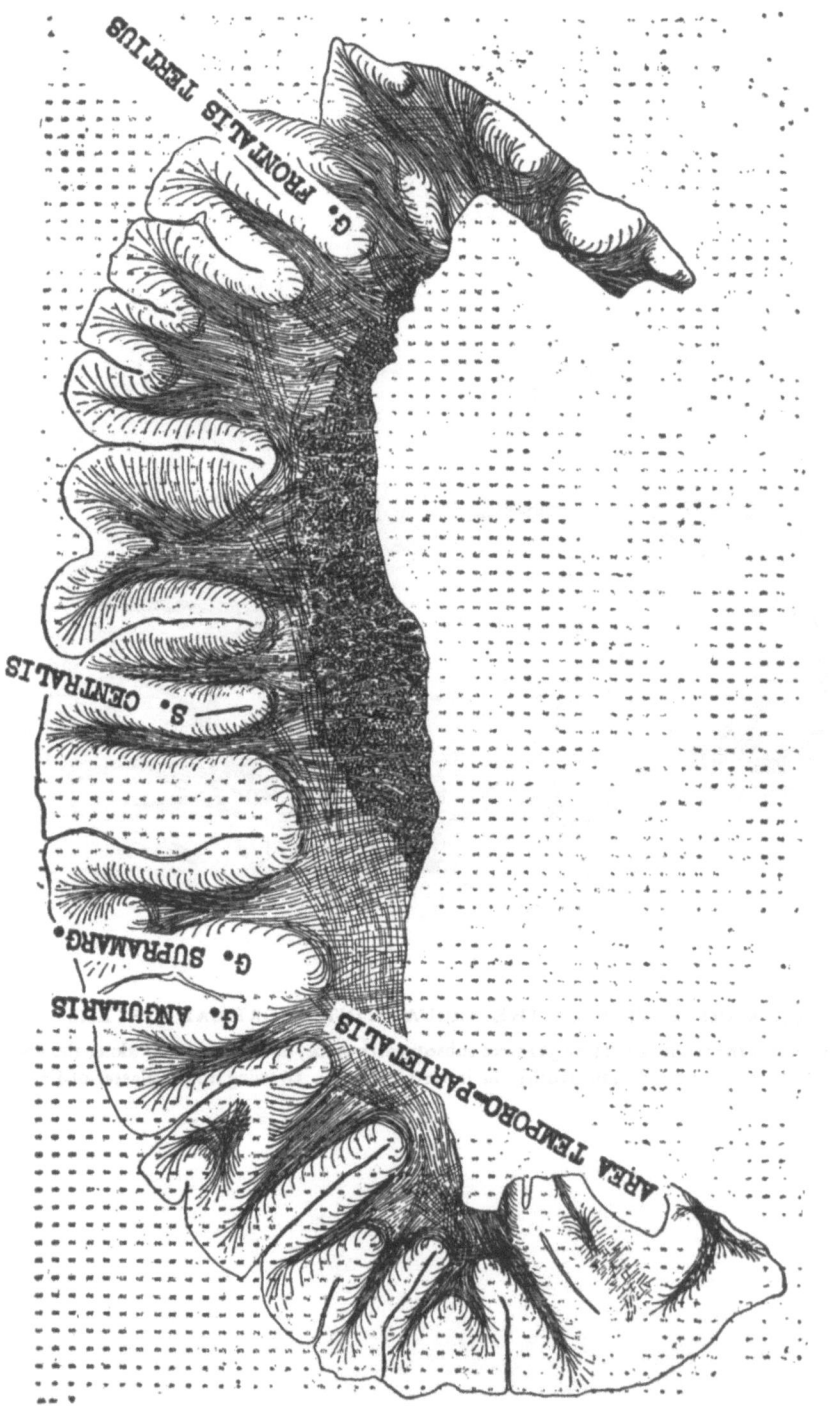

FIG. 36. AN ARCUATE SECTION OF THE ARCUATE FASCICULUS SHOWN IN FIGURE 35 (SEE TEXT), DRAWN AT MAGS. 50 AND 150, WEIGERT-PAL STAIN

The fiber bundles are not sufficiently bent to justify the interpretation that they enter or leave the crests of the convolutions. There relation to the subcortical laminae of the fissures is obvious.

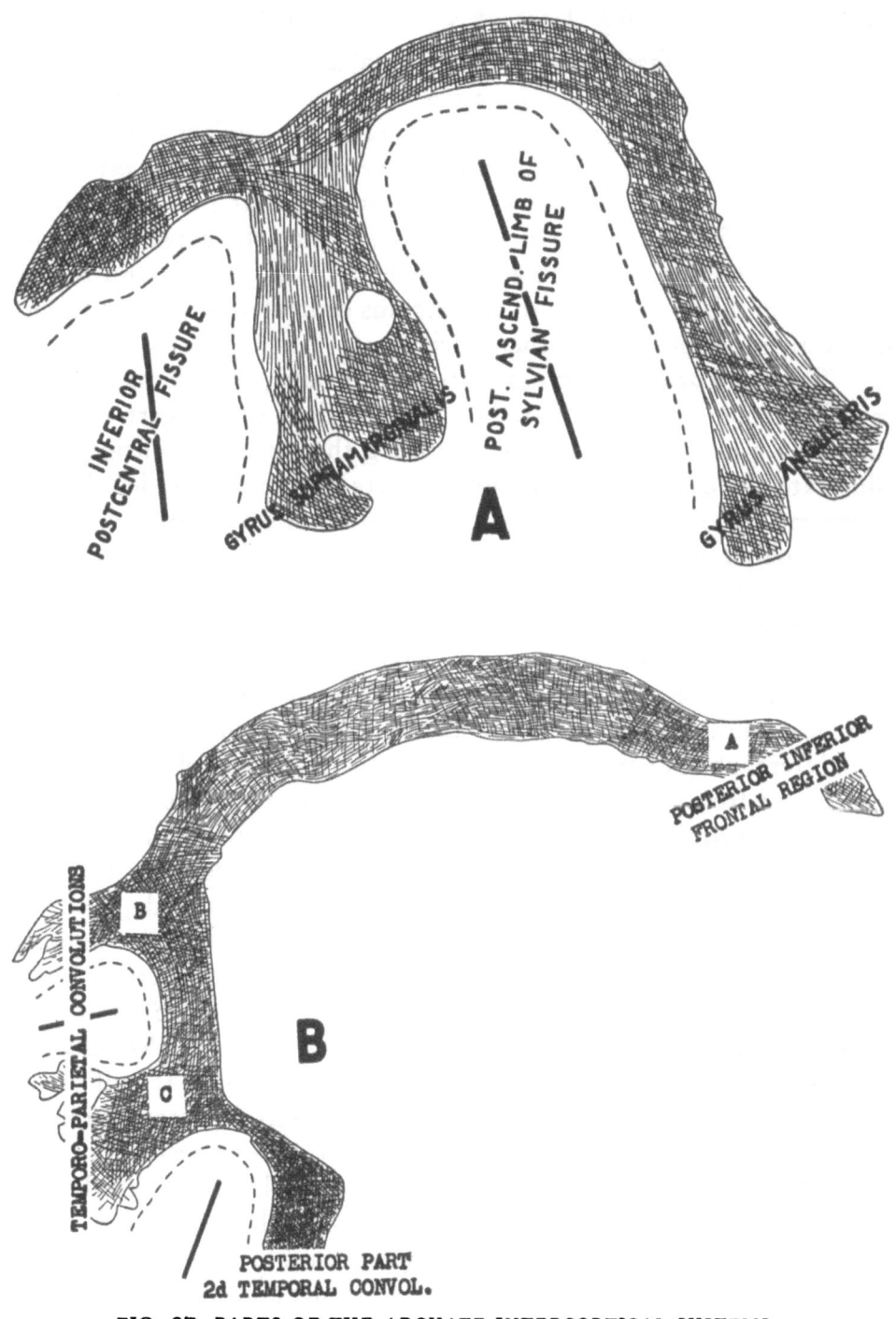

FIG. 37. PARTS OF THE ARCUATE INTERCORTICAL SYSTEMS

A, a fine lamina of the arcuate fasciculus between the postcentral gyrus and the gyri supermarginalis and angularis.

B, a procession of the band between the postero-inferior frontal region, two parieto-temporal gyri and the posterior part of the second temporal gyrus. (See text.)

have described a fasciculus centro-parietalis. Quensel (6) speaks of a bundle which he calls by the names of fasciculus longitudinalis superior or fasciculus centro-parieto-occipitalis. Mayendorf's (1) opinion on the general subject of the arcuate fasciculus is difficult to interpret. He maintains that that fiber system is nothing but the U-system of the Sylvian fissure; that it does not exist; that the capsula externa contains none but the association systems of the insula; and he describes long fibers arranged in a fasciculus centralis obliquus, a fasciculus centro-parietalis and other fasciculi.

The photographs of gross sections of brains, containing areas of massive degeneration prepared by the method of Weigert-Pal, adduced by a number of authors, including the two last mentioned, to illustrate the existence of the nerve tracts in question, failed to convince me. I could see in those photographs nothing beyond a very diffuse pallor on a somewhat darker background. I therefore proceeded as follows.

A hemisphere exploded by the method described was torn apart along the line of cleavage which always forms lateral to the densest mass of the corona radiata. The mesial plane of the arcuate sheet, adherent laterally to the subcortical laminae of the fissures, was exposed. The lateral shell of the hemisphere was then broken below, along the line of the fronto-parietal operculum; in front, along the crest of the precentral convolutions; and behind, irregularly along the posterior parietal region. Manual dissection of the arcuate sheet was then continued until a single band, about seven millimeters wide, was left across the middle of the hull-like subcortical laminae of the fissures, as may be seen in Figure 38 A. The band was then dissected off, exposing to view the hull-like structures (Figure 38 B). The band itself was flattened and cut into microscopic sections. A detailed drawing of one of the sections may be studied in Figure 38 C. It will be observed that continuous fiber bundles extend only between the limits marked "A" and "B." The space to the left of A corresponds to the part of the band which in the dissection appeared to pass into the precentral convolution as far as its crest; that to the right of B corresponds to the part of the band which was insinuated in a convolution of the inferior parietal lobule. Obviously, projection fibers of a radial direction were so interlaced with the sagittally placed arcuate fibers in the bases of the convolutions as to produce the appearance of continuity.

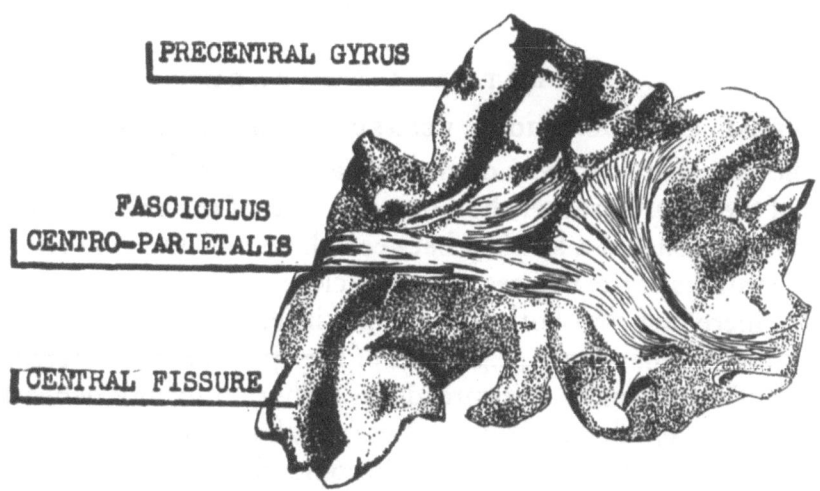

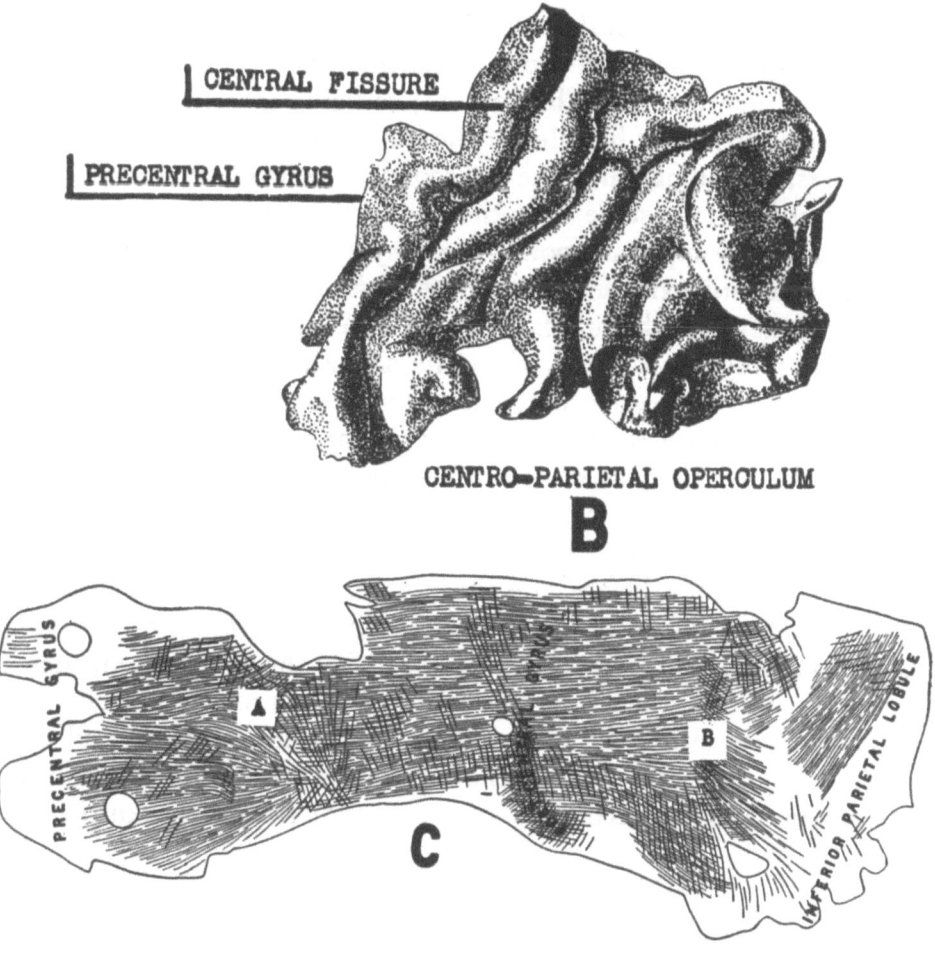

FIG. 38. A DISSECTION OF THE FASCICULUS CENTROPARIETALIS

A and B. The obverse of the configuration of the cortex. Drawings from metal casts of dissections.
C. A section in the plane of the centro-parietal fasciculus shown in Figure 38 A, at mags. 50 and 150.

2. General Considerations regarding the Arcuate Systems

The fact that the fibers of the arcuate band extend between the cortex of the fissures, rather than between that of the crests of the convolutions, establishes a kinship between them and the "short" intercortical systems of the fissures, which is closer than might be assumed from the mere fact that both systems are intercortical. In the smooth brain of the low mammalia, e.g., that of the opossum, the intercortical fibers of different lengths are spread out rather uniformly beneath the cortex (Figure 39). In the course of mammalian evolution the cortex increases in extent and the cerebrum becomes fissured. Underlying the cortex of each fissure is a lamina of intercortical fibers. There is no reason, however, for assuming that the ancient intercortical systems of the smooth or the little-fissured cerebrum entirely disappear and are entirely replaced by the newly-formed intercortical systems contained in the fissures. Numerous analogies bear testimony to the fact that newly-formed cerebral structures are superimposed upon archaic structures without actually replacing them. The irregular network of single fibers mentioned in Chapter III, as being intermixed with the regular aggregations of bundles of the human intercortical systems of the fissures, is strikingly like the intercortical network of the opossum. It is not unsafe to assume that the new intercortical systems of the fissures are superimposed upon ancient systems of like function, but that the latter still exist, and that some of their constituents are buried at a greater depth in the substance of the cerebrum.

If the foregoing be true, then the arcuate system may be considered to represent an intercortical system which, in the order of cerebral evolution, preceded the intercortical systems of the fissures. The fact that its fibers do not terminate in the cortex of the same fissure in which they originate, strengthens such a hypothesis. When any given area of the cortex expands so that many fissures are formed where there were few before, the old intercortical systems, still retaining their former points of origin and termination, must now extend between fissures at greater distances apart. By the same token, such a hypothesis is strengthened by the fact that the arcuate system does not extend into either the occipital or the anterior frontal and prefrontal areas of the cerebrum. If the cortical area which corresponds to the arcuate band has become of greater extent than it was in preced-

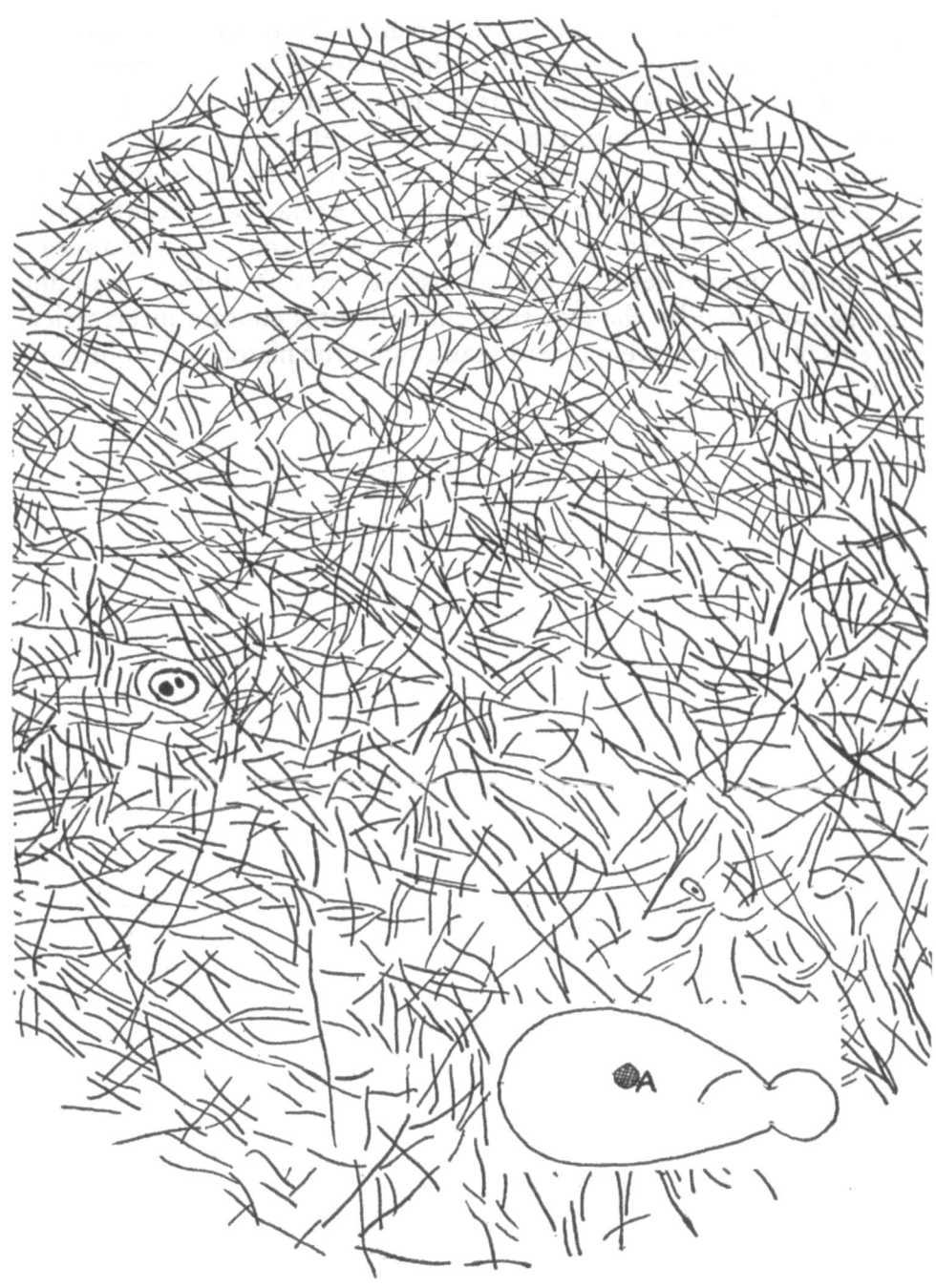

FIG. 39. THE TANGENTIAL IRREGULAR NETWORK OF SINGLE NERVE FIBERS IN THE LOWEST CORTICAL LAYERS OF THE CEREBRUM OF THE OPOSSUM

This network of single fibers is strikingly like the network which is intermixed with the regular fabric of intercortical nerve bundles underlying the cortex of the fissures of the human cerebrum. A close study of the photomicrograph, Figure 10, will reveal the fibers in question (Chapter III).

A, the spot from which the drawing (× 150) was made.

ing stages of cerebral evolution, the span of the band must have become proportionately longer. But the occipital and the prefrontal areas have been added at the ends of the cerebrum, outside the limits of the arcuate system; into those regions therefore the ancient system does not extend; and whatever connections the new polar portions of the cerebrum have with the older portions are relatively recent. It was pointed out in a preceding chapter of this work that the late appearance in the human fetus of certain fissures of very ancient standing may be explained by the fact that those fissures contain intercortical systems of comparatively recent standing in the history of evolution.

CHAPTER TWELVE

INTERCORTICAL SYSTEMS OF THE LATERAL TEMPORAL AREA

1. The Superior Temporal Fissure

In the map (Figure 40), where the intercortical pathways which traverse the fissures accentuate the parts played by the smallest offshoots and by the annectants, the superior temporal fissure is discovered to be the most complicated of the entire fissural system of this hemisphere.

Orientation.—Three prominent fissures traversed the lateral surface of the temporal lobe of the hemisphere mapped. The course of the highest was horizontal; that of the middle one was horizontal in front, becoming somewhat diagonal as it passed backwards and downward to the base (Figure 20), where, after proceeding for some distance in the temporo-occipital region, it curved laterally again, and terminated on the lateral suface near the margin. The course of the lowest of the three fissures was downwards and backwards from the lateral surface, on to the base, where it terminated near the margin.

From above downwards, the anterior termination of the three fissures was successively farther forward, so that the lowest of the three limited the region of the temporal pole. A number of small sulci, which communicated with these fissures in front, operated to complicate the simple arrangement.

Between the anterior portions of the superior and middle temporal fissures there intervened a horizontal sulcus, about two centimeters long, which had at about its middle one offshoot above and one below. These latter joined respectively the superior temporal fissure above and the middle one below. The anterior termination of the superior temporal fissure was immediately in front of that junction; that of the middle one terminated about three centimeters farther forward in a vertical bifurcation. Immediately opposite its junction with the cruciate sulcus, the lower wall of the middle temporal fissure had a vertical offshoot which terminated below in a horizontal bifurcation. About a centimeter back of that, it was partly severed by an annectant convolution.

Such was the appearance of this plexus of fissures, as viewed from the side of the cortex. The aspect of the observe side of the cortex—that of the hulls of the boatlike structures—was much simpler (Figure 4). The superior temporal fissure, the entire intermediate cruciate sulcus, and the entire anterior portion of the middle temporal, as far back as its anterior annectant, were embraced in a single hull or half cylinder, which was entirely separated from the rest of the middle temporal fissure. As has been explained, the method employed consisted in preparing microscopic sections from such hulls or half cylinders after they had been flattened. In the enumeration of the pathways, therefore, the anterior portion of the middle temporal and the entire cruciate sulcus above it are included in the superior temporal fissure system. Although such an arrangement is rather unnatural from the point of view of surface topography, it is entirely natural from the point of view of the underlying fiber systems.

The anterior portion of the middle temporal and the horizontal limb of the superior temporal fissure together contained seven annectant gyri which rose in regular succession alternately from the upper and the lower walls of the fissure. The obverse of these annectants may be seen in Figure 4 as linear depressions, giving the half cylinder a spiral appearance. In the description of the pathways, these annectants will be signified by their numbers, beginning with the most anterior.

Posteriorly, the horizontal portion of the superior temporal fissure described a sharp curve upwards and ascended the parietal lobe. At the point of the angle formed by the horizontal and ascending portions, the fissure had a short offshoot postero-inferiorly; antero-superiorly there proceeded from the angle a long sulcus which entered the posterior portion of the Sylvian fissure. The temporal fissure was partly separated from this communicating sulcus by an annectant gyrus. The ascending limb of the fissure had two offshoots; one about the middle of its posterior border, and another at its highest part, in front of a superior annectant which separated this fissure from a higher portion of the temporo-parietal fissure proper.

Pathways of the superior temporal fissure.—1. Between the anterior bifurcated extremity and (a), the posterior extremity of the intermediate horizontal sulcus; (b), the annectant which separates the conglomerate anterior portion of the fissure from the middle temporal; (c), the bifurcated antero-inferior offshoot.

2. Between the superior border of the horizontal intermediate sulcus and (a), the annectant which separates the anterior conglomerate extremity of the fissure from the middle temporal; (b), the posterior half of the inferior border of the horizontal intermediate sulcus.

Posterior to these pathways and parallel to them, fiber bundles may be seen to cross the fissure diagonally in the horizontal part of its course.

3. Between the inferior border of the antero-inferior bifurcated offshoot and a small portion of the superior border of the superior temporal fissure proper near its extremity.

4. More or less vertical (as viewed on the map) pathways in the anterior bifurcated extremity.

5. Between a small part of the inferior border of the fissure immediately behind the horizontal intermediate sulcus and (a), the posterior border of the postero-inferior offshoot; (b) the postero-superior offshoot.

6. Between the annectant which separates this fissure from the communicating sulcus with the Sylvian fissure and the inferior border, posterior to the fourth incomplete annectant.

7. Vertical pathways in the posterior-inferior offshoot.

The following pathways have been mapped in the ascending portion of the fissure:

8. Between the inferior border of the postero-superior offshoot and (a), the antero-superior offshoot; (b), the superior annectant which separates this fissure from the temporo-parietal proper.

9. Pathways between the opposite borders around this ascending portion of the fissure.

2. THE MIDDLE TEMPORAL FISSURE

Orientation.—In this hemisphere the middle temporal fissure extended partly on the lateral surface of the temporal lobe, partly on its base. In front it was united with the superior temporal by the interposition of a complicated plexus of sulci, described as part of the superior temporal. On the base, the posterior portion of the middle temporal fissure passed backwards for some distance, then curved laterally, terminating at right angles to the inferior lateral occipital fissure from which it was separated by a thin convolution. It had three annectants in its course; an anterior one separated this fissure from the anterior conglomeration of sulci; a posterior was in the basal course of the fissure; a third was midway between the two.

Pathways.—1. Between the anterior annectant and (a), the middle annectant; (b), the upper border of the fissure in the space between the anterior and middle annectants.

2. In the space between the middle and posterior annectants two sets of pathways may be seen on the maps (Figures 20 and 40) extending more or less diagonally across the fissure in opposite directions.

3. In the last part of its course, between the posterior annectant on the base of the hemisphere and its termination on the lateral surface, the pathways of this fissure are extremely complicated. At about the middle third of this portion of the fissure, they may be seen to extend squarely around it. Anteriorly, fiber bundles extend parallel to the posterior annectant around the fissure. More mesially fiber bundles radiate between the anterior border and (a) the lateral border; (b), the floor of the fissure; (c), the mesial border. More or less diagonal pathways extend around the fissure between points on the lateral border and more posteriorly situated points on the mesial border. In the sagittal course of this portion of the fissure pathways may be seen on the map (Figure 20) to extend in its long axis between an anterior offshoot and the posterior portion of the lateral border of the fissure. In the terminal transverse course of the fissure pathways extend between the somewhat bifurcated extremity and the posterior half of the mesial border of the basal portion of the fissure.

CHAPTER THIRTEEN

INTERCORTICAL SYSTEMS OF THE LATERAL OCCIPITAL AREA (FIGURE 41)

1. THE SUPERIOR LATERAL OCCIPITAL FISSURE

Orientation.—At the junction of the anterior and middle thirds of its course, this fissure anastomosed on the surface, by means of a narrow communicating sulcus, with the temporo-parietal fissure, but was separated from it by an annectant gyrus placed in the opening of the communicating sulcus. A few millimeters behind this annectant, an irregularly conical projection arose from the floor of the fissure. At the junction of the middle and posterior thirds of the fissure, there was another conical projection on the floor of the fissure. Behind the posterior conical annectant, the fissure was in the form of an irregular letter U, into the mouth of which was received the sulcus occipitalis transversus. The right limb of the U was bifrucated above; the left terminated above without any bifurcation. The horizontal piece of the U was biconcave and it extended both to the right and the left beyond the vertical limbs, in the form of two inferior offshoots. In the antero-inferior part of its course, the fissure had two small offshoots: an anterior one immediately above the inferior incomplete annectant and a posterior one between the two conical annectants. An incomplete annectant rose from the upper wall of the horizontal piece of the U-like posterior portion of the fissure.

Pathways.—1. Between the annectant which separates this fissure from the inferior lateral occipital and (a), the antero-superior border; (b), the incomplete annectant which separates this fissure from the temporo-occipital fissure.

2. Between the anterior conical annectant and (a), the antero-inferior partial annectant which separates this fissure from the temporo-occipital; (b), the adjoining small part of the anterior border; (c) the posterior part of the lower border of the temporo-occipital fissure.

3. In the part of the fissure behind the anterior conical annectant:

Two sets of diagonal pathways between the opposite borders of the fissure which cross each other at right angles.

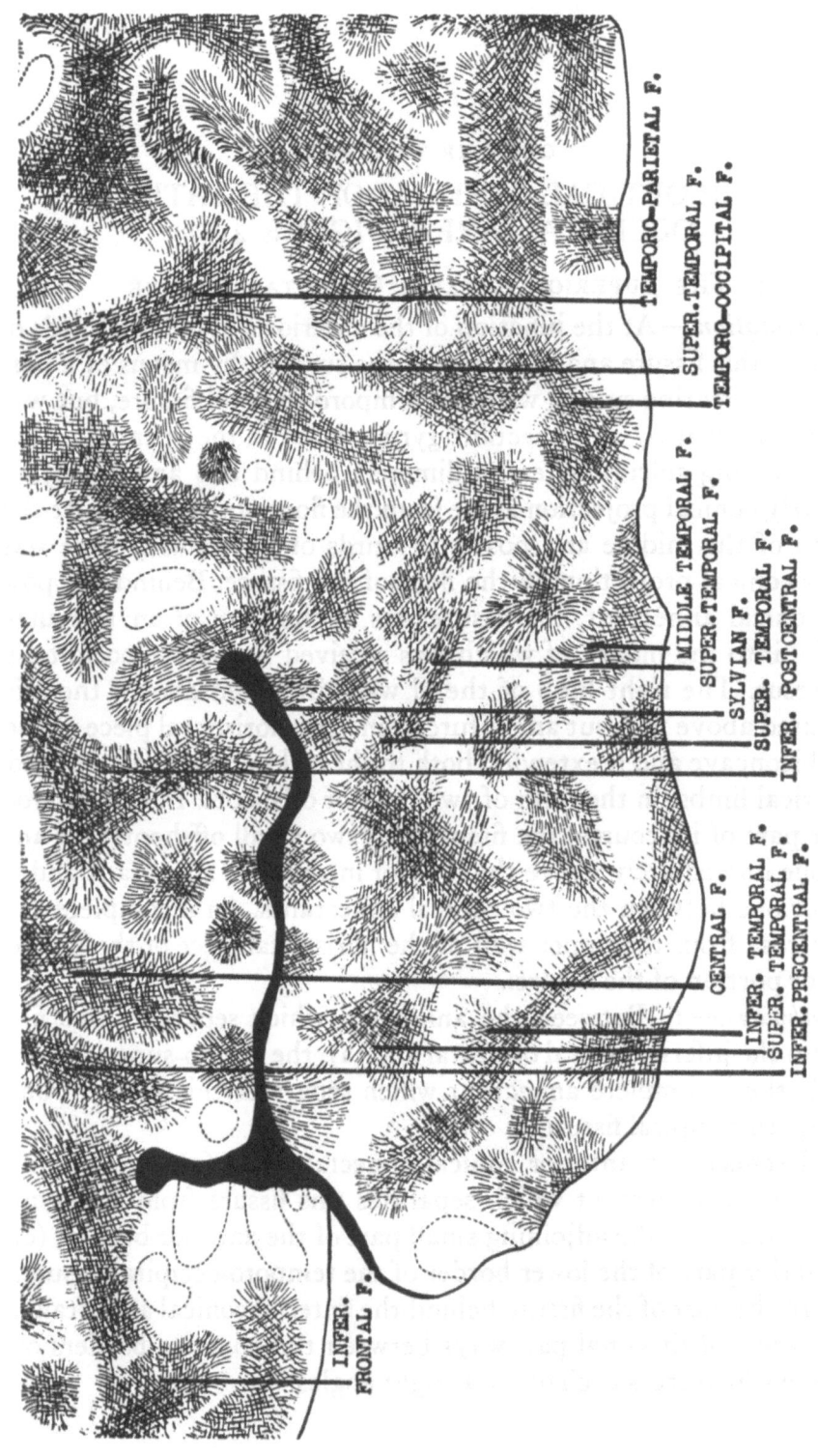

FIG. 40. A COMPOSITE SCHEMA OF THE INTERCORTICAL SYSTEMS OF THE LEFT LATERAL TEMPORAL AREA OF A HUMAN CEREBRUM

4. Vertical pathways in the postero-superior offshoot.

5. In two situations there are pathways which extend transversely around the fissure; (a), in the antero-inferior portion, below and in front of the anterior conical annectant; and (b), in the posterior portion, which lies between the two superior and the two inferior offshoots.

6. Between the anterior border of the postero-superior offshoot and the posterior limb of its bifurcation.

7. Between the antero-inferior offshoot and the middle third of the inferior border of the fissure.

2. The Inferior Lateral Occipital Fissure

Orientation.—In this hemisphere the inferior lateral occipital fissure, situated near the lower border, was, in the greater part of its course, horizontal. An annectant gyrus separated it from the junction of the superior lateral occipital and the temporo-occipital fissures. On the surface, the linear arrangement of the conglomerate of small sulci of the posterior termination of this fissure, was somewhat in the form of an irregular =(, but the result of the obliteration of the angles brought about by opening the fissures into wide grooves, as represented on the map, is that the terminal piece assumed the form of a trifolium. The latter was partly separated from the main fissure by an incomplete annectant which rose from its upper wall. The superior leaf of the trifolium was directed vertically upwards; its inferior leaf, forwards; and its central leaf at first backwards, then, turning on the mesial surface of the occipital lobe, for a short distance upwards, towards the pole. The horizontal part of the fissure had two small offshoots; one anteriorly, a short distance behind the anterior complete annectant; and one posteriorly in front of the posterior incomplete annectant.

Pathways.—1. Between points on the inferior border and more posteriorly situated points on the superior border. In the anterior part of the fissure these pathways extend between the anterior annectant and the superior border. In the posterior portion they extend between the inferior border and the posterior border of the superior leaf of the trifolium.

2. Between points on the inferior border and more anteriorly situated points on the superior border. In the anterior part of the fissure

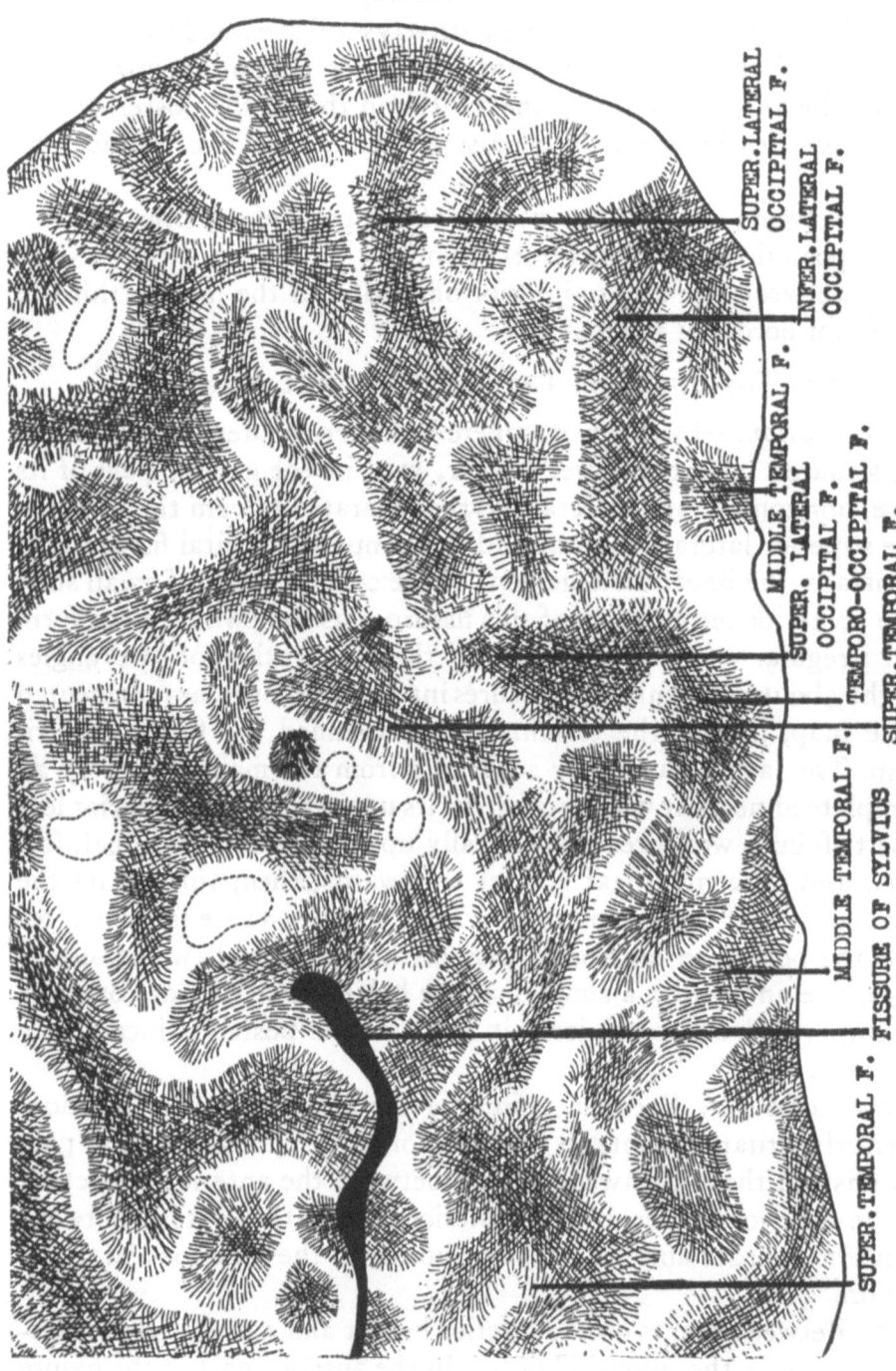

FIG. 41. A COMPOSITE SCHEMA OF THE INTERCORTICAL SYSTEMS OF THE LEFT LATERAL OCCIPITAL AREA OF A HUMAN CEREBRUM

these pathways extend between the inferior border and the anterior annectant. In the posterior part of the fissure they extend between the superior border and the central leaf of the trifolium.

3. Pathways in the form of a narrow band around the body of the trifoliate termination.

4. Between the inferior border of the inferior leaf and the opposite border of the body of the trifolium.

5. Between the inferior and central leaves of the trifolium.

6. Between the superior and central leaves of the trifolium.

7. Between the superior and inferior borders of the inferior leaf of the trifolium.

3. The Temporo-occipital Fissure

Orientation.—This fissure had two inferior offshoots, one near its anterior termination and the other at the point of the angle where the fissure turns in a direction upwards and backwards to join the two lateral occipital fissures. In the middle part of its course it was interrupted by an annectant gyrus.

Pathways.—1. Between the anterior termination and (a), the partial annectant which separates this fissure from the superior lateral occipital; and (b), a part of the inferior border back of the angle of the fissure.

2. Between a part of the inferior border in front of the angle of the fissure and the lower half of the posterior annectant which separates this fissure from the inferior lateral occipital.

3. In a radiating manner between an antero-inferior offshoot and the anterior half of the superior border.

4. Between the opposite borders of the fissure. In the situation of the postero-superior partial annectant these pathways extend between that annectant and the inferior border, including the inferior angle.

CHAPTER FOURTEEN
GENERAL CONSIDERATIONS

1. The Pattern of the Subcortical Pathways of the Fissures and Its Complications

A study of the composite pictures shows that a number of the smaller sulci and depressions, which do not communicate with others, are covered by a crosshatching which represents two sets of parallel bundles crossing each other. Without any reference to them as connecting links between different cell areas of the cortex, such an arrangement of the intercortical pathways of the fissures may be taken on the whole to be the basic pattern. From the superficial viewpoint of the gross morphology of the cortex, which unfortunately must be adhered to until more data on the subject are available, the reason why most of the fissures contain more than two sets of pathways is apparently because of their anastomoses. Thus when two fissures are joined, for instance, at a right angle, the transverse bundles of each are more or less parallel to the longitudinal bundles of the other, so that no additional sets of pathways are superimposed on the basic two sets at the point of junction. But if the anastomosis is at an angle considerably more or less than 90 degrees, the transverse bundles of each of the fissures may cross both the transverse and the longitudinal bundles of the other, so that two more sets may be superimposed at the point of junction. Small offshoots may thus give rise to complicated patterns in the bed of the parent fissure. And as each of the larger fissures has a number of tributaries, small offshoots and terminal bifurcations, and anastomoses besides with other large fissures, the simple basic pattern of pathways becomes complicated by the addition of similar patterns.

If the angle of anastomosis of two fissures is very obtuse, then the long pathways of each accommodate themselves to the course of those of the other. Whether the longer procession of the pathways of one fissure in the bed of another has any bearing on the phylogenetic and ontogenetic degree of relationship of the two anastomosing fissures, must be left to the decision of the morphologist and of the cell architect.

Considering the prevailing tendency of the intercortical pathways to a direct course from convolution to convoluion, the curves which may be observed in a number of situations, particularly near the junction of fissures, are probably distortions brought about by the unequal expansion of the different parts of the cerebrum during the period of growth. The hairpin curves in fissures of certain types of of cerebrum, such as those seen in Figures 27 and 28, are difficult to explain. They remind one of the hairpin curve formed in the substance of the brain stem by the seventh nerve and by the occasional fasciculus circumolivaris pyramidis. It is possible that these strongly curved intercortical pathways are straight to start with, connecting points at short distances from each other, and that they are pushed down by the growth of the tissue on one side of them, such as protruding buttresses.

Further complications of the pattern of the intercortical pathways of the fissures result from the fact that the fibers of the deepest layers of the lamina in which these pathways are contained are longer than those situated more superficially. While fibers begin or end at numerous points throughout the cortex of a sulcus, no massive interruption can be observed except in the situation of the annectant convolutions. And although the course of many of these fiber bundles is continued beneath the bases of the annectants, most of them may be observed to rise to or descend from the annectants in the same way that they ascend to or descend from the surface convolutions. A definite linear hiatus is thus produced along the line of an annectant in the field of these fibers.

Instances in which entire fiber sheets extend on the deep side of cerebral areas of large extent were shown in the case of the insula and its opercula and in the case of the parieto-occipital and calcarine fissures. The arcuate fasciculus or sheet and the inferior longitudinal fasciculus must be included in the same category. These deep fiber sheets are probably the vestiges of an ancient intercortical system of the smooth or the slightly fissured cerebrum. A further complication of the general pattern is brought about by the fact that these "long association" fibers, as well as a number of projection fibers, perforate the subcortical lamina of the fissure to enter (or to leave) its overlying cortex. How far such fibers travel in the plane of the lamina before attaining to (or emerging from) the cortex I do not know, but that they do proceed in it for a variable distance may be concluded with certainty from the microscopic preparations shown in Figures 36 and 37. In the dissection of the intercortical lamina of a fissure, the deepest part of the floor may be seen in some instances to be anchored to the central mass of the cerebrum, like a sailboat with the keel grounded in sand. This is particularly true in the case of the central fissure (Figure 36). Dissection along the line of this "keel," when it exists, is, therefore, always difficult. Microscopic preparations of cross sections of such a boatlike structure with its keel of fiber bundles, show, however, that the number of radial fasciculi which perforate the intercortical lamina is small compared with that which insinuates itself between each two of the curved laminae to reach the crest of the convolution. In the sagittal sections of the automatically produced gross lines of cerebral cleavage (Figure 1), the former cannot be found while the latter appear with great clearness. Minkowski (51) has shown that when the crest of a convolution is experimentally injured the adjoining tissue of the wall of the fissure remains intact. His conclusion is that the crest of the convolution is related *en preferance* to the long fibers—projection and long-association fibers. Mayendorf's (1) rather severe criticism of Dejerine's (19) statement that the projection fibers are largely related to the crests of the convolutions is not justified by the evidence of either gross or microscopic preparations.

2. Criticism of Methods

From the number and the degree of divergence of the subcortical pathways of the fissures, as well as from the fact that they are made up of fibers of very different lengths, it will easily be concluded that tracts of degeneration must rapidly fade out along a number of radii drawn from a cortical lesion as a center. The current method of cross-sectioning the tissue, which even in very favorable locations of the cerebrum produces a picture consisting either of fine stippling or of minute streaks, can give no idea of the course of the intercortical pathways. Even in Marchi degeneration preparations of animal brains, it must be a matter of extreme difficulty to distinguish the black granules and to align them from section to section in such a way as to reconstruct even a semblance of the order in which the bundles are arranged in the complicated fabric. In the ordinary Weigert-Pal preparations of massive degeneration, even of such excellence as those by which Probst and Quensel studied a number of the long cerebral fiber systems, all that one sees is a diffuse pallor in a surrounding blue field. These authors, however, cautious in their conclusions even with regard to the simpler "long association systems," almost entirely ignore the subcortical systems of the fissures. But Mayendorf, whose Weigert-Pal degeneration preparations (as far as can be judged from his illustrations) are not as good as Probst's or Quensel's, is far more certain than they are of his conclusions. How he was able to arrive at any conclusions regarding the course of the intercortical pathways of the fissures is not, however, made clear either from the nature of the method employed or even from the statements of the author. This is all the more surprising in view of that author's thorough consciousness of the inadequacy of the current methods for that kind of investigation, in view of his superb critical ability, and in view of the exacting condition of "the beginning and ending of the fibers in the cortex," which he justly lays down as the prerequisit of any surmise that certain nerve fibers are intercortical fibers.

3. A Contribution to the Knowledge of the Causes of Fissuration

In view of the fact that systems of intercortical pathways underlie the cortex of every fissure in all its ramifications, it becomes both interesting and important to ascertain the anatomical significance of such universal association. Two possibilities present themselves. One is that the fissures in some way favor the aggregation of the intercortical pathways. The other is that the existence of these pathways, or of their precursors, is productive of those differences in the direction of the lines of growth which result in fissuration.

The first assumption is not so far-fetched as it appears to be at first sight. Elliot Smith (35), (52), thought that sulci were generally formed along lines which separate areas of different cell structure. As far as I can make out, the immediate cause of fissuration is, according to him, a purely mechanical one; the different expansive pressures during the period of growth of, or on, the cell areas of different consistencies. A good example of the extent to which Smith has been supported in his hypothesis of the function of the fissures as "sulci limitantes," is Jefferson's (38) excellent study of the interparietal fissure. In pointing out what he considered as errors made by preceding students of that fissure with respect to the homology of certain of its parts, that author argues that the fissure in question was named *before the precise method of the formation of sulci was known;* that the principal factor in the process of fissure formation is to be found in the different areas of cortical cell structure. The postcentral rami of the interparietal fissure, therefore, owe their existence to the line of separation between the area of common sensibility and the association area of Flechsig; the ramus occipitalis owes its existence to the fact that it marks the line which limits the peristriate area; and so forth.

In the light of the present study, the foregoing hypothesis gains indeed in strength. For if the fissures be *sulci limitantes* between different structural and functional cortical areas, containing as they do intercortical connections of vast extent, they are at the same time *loci conjugentes* of these areas—which might naturally be expected. But the hypothesis does not entirely coincide with a number of facts and has, therefore, been modified by later investigators. A glance at the pictures of the cortical cell areas outlined by Elliot Smith himself (35), (52), shows that certain fissures lie entirely in the midst of areas of uniform cell structure, such as the calcarine fissure in the midst of the area striata, while the lines which separate other areas pass along or across crests of convolutions. An example of a convolution which limits areas of different cell structure is the prominent annectant in the upper half of the central fissure. "It is evident," says Campbell (47), that the majority of the cells (giant cells of Betz), certainly more than half, reside above the level of the annectant buttress; the barrenness of the buttress is clearly indicated; then after a long stretch, also relatively poor in cells, it is seen how they rapidly disappear as the lower level of the fissure is approached." Brodmann's work (53) shows nearly throughout a discrepancy between the lines of the sulci and those bordering the different cell areas. A good example is the paracentral lobule (54). The giant cells of Betz are, according to that author, in front of a line drawn on the lobule in the direction of the highest part of the central fissure, thus placing the paracentral sulcus entirely within the area of giant cells. O. Vogt's position regarding the question in hand is not readily understood. In 1903-4 (55) he wrote as follows: "At present we can only say that a general topographic orientation demands a division of the cerebral cortex on the basis of its furrows; such a division has, however, no physiological value." Throughout his numerous succeeding works (some of them in collaboration with C. Vogt), his opinion is modified. Thus in 1922–23 (56), while still maintaining his opinion that the fissures "furnish partly none partly only a very crude idea of the lines bordering different cell areas," he agrees that the latter lines do have something to do with the *formation* of the fissures. Economo (57), having delimited some 120 cytoarchitectonic areas (58), came to the conclusion that there exists a distinct functional difference between the crest of a convolution and a fissure. He called attention to the fact that the cortex is thickest on the crests of the convolutions, thinner on the walls of the sulci and thinnest in their floor. He showed, moreover, that the difference in the total thickness of the cortex in these several parts is due entirely to the fifth and sixth cell layers. And hypothetical though his opinion appears to be, that the crests of the convolutions are, therefore, mainly efferent and the sulci mainly afferent in their function, it deserves attention. It was pointed out above, that the present work is corroborative of Dejerine's and others' opinion that the projection fibers are to be found "en preferance" between the boat-like laminae which contain the intercortical systems of the fissures, that is, in the crests of the convolutions. To this extent Economo's hypothesis is strengthened by the present work.

Kappers (59) argued for a close relationship between the cortical cell areas and the fissures, from a phylogenetic point of view. He showed that in the history of the evolution of the insular and frontal regions, the changes in the cell areas are closely connected with corresponding changes in the pattern of fissuration. And he was impressed by the fact (60) that the changes in the pattern of the fissures, from the point of time, tarry behind those of the cell areas, thus accommodating themselves to the latter changes.

The question regarding the underlying or general cause of fissuration is old. Turner (61) ascribed it to the resistance offered by the skull capsule. He ascribed the cause of a particular pattern of fissuration to the shape of the skull, the convolutions being formed at right angles to the lines of pressure. But Schnopfhagen (21) examined seventy human and a large number of animal embryos with that point in view. He found that after the pattern of the fissures had been completed, the skull capsule was not yet quite filled. He concluded that fissuration was due to the greater growth energy of the cerebrum along the lines of the convolutions, whereby the spaces between them were left behind as sulci—an explanation which is obviously too general and therefore no explanation at all.

That a temporary disproportion between the growth of the cranial cavity and that of its contents may be the cause of the transitory fissures in the young human fetus, is not improbable. It appears that the temporary furrows are coincident with the formation of the occipital lobes. Cunningham's argument (14) to that effect is worth quoting:

> We may assume [says he] that although cranial and brain growth, as a rule, go on smoothly and evenly and in perfect harmony with each other, all steps toward an advance in development must be initiated within the brain and that, for a time at least, the enclosing skull-capsule will resist these. This being granted, we can readily understand that the tendency toward the cerebral growth, which gives rise to a well mapped out occipital lobe, is more firmly impressed upon the brain than upon the skull. When the primate head reaches in its development the quadruped stage, the cerebrum goes on without any intermission in its growth, towards the higher development and the formation of a distinct occipital lobe. The cranium, however, pauses in its growth. But this quadruped pause marks only a stage in its evolution; it is merely temporary, although it is of sufficient duration to produce the infoldings of the cerebral wall.

Although the actual relation of cause and effect in the foregoing argument is beyond proof, the assumption itself is not in opposition to any known facts, and its logic carries conviction.

The case is entirely different with respect to the cause of permanent fissuration. I again quote Cunningham (62):

> That the initial cause of the formation of the cerebral furrows is to be sought for in a restraint upon the surface-growth of the cerebrum, I would deny in the strongest terms. At the time when the first permanent cortical fissures appear, the cerebrum lies loose within the cranial capsule and does not completely fill it.... It is a fact which I have had many opportunities of verifying.... I do not believe that under normal conditions the directions of the furrows and convolutions are affected by a restraint placed upon the growth of the cerebrum by the skull-capsule.... If the theory of cerebral localisation be true, the cortex must grow in accordance with the functional activity which is displayed in its different parts.

Retzius (12) was equally opposed to the idea of a restraint by the cranial capsule as a cause of fissuration. He maintained that whatever the cause of fissuration may be, it is not of an external nature, but is inherent in the forces of tissue growth.

Baillarger's (63) explanation of the causes of fissuration has found almost universal support. It is a fact that the volumes of similar bodies are related to each other as the cubes of their diameters, while their surfaces are related as the squares of their diameters. The surface of the growing cerebrum thus being unable to contain its volume, wrinkles up. Jelgersma (64) proposed this hypothesis of the general cause of fissuration without apparently being aware that it had been previously promulgated by Baillarger. Schnopfhagen, Cunningham, Retzius and a number of other investigators speak of it in terms of approval.

That this physical law is not only inadequate to account for any particular pattern of fissuration, but is of itself insufficient even as a general cause of that phenomenon, may be readily gathered from a few familiar examples: The cerebellum is much smaller than the cerebrum, yet is is much more fissured. In small animals, with a smooth cerebrum, the still smaller cerebellum is fissured. The kernel of the large cocoanut is smooth, that of the walnut or hickory nut is very much convoluted. Certain huge varieties of melons are perfectly smooth; others, of a much smaller size, are thoroughy wrinked. Such examples might be multiplied at great length. The fact that one organism or part of an organism is enabled to exist and to maintain its particular form by virtue of a given physical law does not mean that another organism, or part of an organism, differently situated, could not maintain a similar form in spite of the operation of the same law. The baker overcomes the operation of the law in question; he renders the surface of the loaf more extensible by means of wetting it. The difficulty due to the discrepancy between the increase of volume and that of surface in growing living bodies is in very many instances overcome by a degree of proliferation of the surface cells in proportion to the degree of the stretching force to which the surface is subjected by the expansion of the contained volume. In other instances, where tearing of the surface is followed by rapid healing, such as is frequently the case in giant puffballs or the outer bark of trees, such tearing is the normal state of things.

It is therefore reasonable to assume that if in the general economy of the species a large and smooth cerebrum had not been a disadvantage, a means for coping with the law of discrepancy between the growing surface and volume would have been provided by the forces operative in the course of evolution. That from an economic point of view such a cerebrum would be a great disadvantage is certain. The huge and delicate mass would then have to be protected by a very thick bony covering; the bony skeleton, in order to support the great weight, would have to be very much thicker and stronger than it now is, and so would the tendons and muscles, the whole necessitating a very much increased apparatus of nutrition.

As a matter of fact, a closer study of the plan on which the cerebrum is built reveals it to be largely a surface organ—one whose parts are laid out on a plane rather than distributed in the volume, as is the case in an organ like the liver. Relatively few fibers are placed radially, relatively many are arranged in the form of a shell near the surface. The growth of the cerebrum is therefore largely an extension of its surface.

Moreover, in the large cerebrum the tangential fibers, which form the shell, are grouped in well delimited laminae, with open spaces between them. Although the radial fibers perforate these laminae at many points, their greatest aggregation is along the spaces between the tangential laminae. During growth the radial lines of force must tend to push the tangential lamina as a whole outwards; the resistance to these lines of force must be greatest along the axes of the lamina, diminishing towards the edges and becoming least in the spaces between the laminae. On the other hand, the resistance met by the lines of force implied by the expansion of the tangential lamina itself, must be greatest at its growing edges. The result of the action of two forces at right angles to each other, radially and tangentially, must be the assumption of a new direction by the combined lines of force, along the diagonal of their parallelogram. The edges of the lamina as they grow, must, therefore, deviate from a tangential towards a radial direction; and as each degree of such deviation must be in each succeeding stage of growth in favor of an added degree of deviation in a radial direction, the ultimate result must be the formation of grooves with perpendicular or nearly perpendicular walls.

Such a probability explains certain hitherto unexplained points and reconciles the different views held regarding the relation of the fissures to the cell areas of the cortex.

Pansch (65) formulated the following simple rule regarding the relation of the order of the appearance of fissures in the fetus to their ultimate depth. Fissures which are earliest to appear are ultimately the deepest. If the foregoing causes of fissuration be true, the reason for that rule becomes simple: The wider the intercortical tangential lamina, the sooner must its deformation become apparent and the deeper must be the fissure which it will form.

It has already been pointed out that Kappers was led by his investigations to the conclusion that a particular pattern of fissuration follows, in the course of evolution, particular cell distributions in different cortical areas. If the causes of fissuration as proposed in the foregoing be true, the cause for the facts observed by Kappers becomes understandable. The tangential laminae, being made up of intercortical connecting fibers, must accommodate themselves in their shape to the arrangement of the cortical cells; and since a fissure forms along the axes of that lamina, a particular pattern of fissuration must follow a particular distribution of cells in the cortex.

The different views held by Elliot Smith, by O. Vogt, by Brodmann, and their several schools may be reconciled, as well, on the basis of the foregoing exposition. The different lengths of the fibers and pathways of the intercortical systems of the fissures speak for the fact that they establish connections (whose nature is still unknown) between cortical cells or groups of cells at different distances apart. There is no reason whatever to assume that they do not serve as connections between cells of the same cortical cell area, as well as between cells or cell groups situated in different areas. The subcortical lamina of a fissure may, therefore, be placed either entirely within an area of uniform cell structure, or along the line which separates two such areas, or near this line, depending upon the functional interrelationships of cells, cell groups or cell areas in the particular area of the cerebrum.

REFERENCES

1. V. Niessl—Mayendorf, E., Die Assoziationssysteme des menschlichen Vorderhirns. *Arch. f. Anat. u. Physiol., Physiol. Abt.*, (1919), p. 283.
2. Arnold, F., Handbuch der Anatomie des Menschen. Freiburg, (1851), II, 721.
3. Meynert, Th., Der Bau der Grosshirnrinde und seine örtlichen Verschiedenheiten, etc. Separatabdr. aus der *Vierteljahrschr. f. Psychiat.* Neuwied u. Leipzig, 1872.
4. Vulpius, O., Ueber die Entwicklung und Ausbreitung der Tangentialfasern. *Arch. f. Psychiat.*, XXIII (1891-92), 775.
5. Poljak, S., An Experimental Study of the Association, Callosal and Projection Fibers of the Cerebral Cortex of the Cat. *J. Comp. Neurol.*, XLIV (1927-28), 197.
6. Quensel, F., Beiträge zur Kenntnis der Grosshirnfaserung. *Monatschr. f. Psychol. u. Neurol.*, XX (1906), 36, 166, 266, 353.
7. Rosett, J., A Study of the Cerebral Fibre Systems, etc. *Brain*, XLV (1922), 357.
8. Rosett, J., A New Anatomic Method, etc. *Transactions* of the A. N. A., 1931, p. 425.
9. Ramon y Cajal, S., Studien über die Hirnrinde des Menschen, German translation by J. Bresler. Part 4 (1900), p. 95.
10. Ramon y Cajal, S., Studien über die Hirnrinde des Menschen, German translation by J. Bresler. Part 5 (1900), p. 5.
11. Heschl, R. L., Tiefen-Windungen des menschlichen Grosshirns etc. *Wiener Med. Wochenschr.*, XXVII (1877), 984.
12. Retzius, M. Gustav, Das Menschenhirn. Stockholm, 1896.
13. Eberstaller, O., Das Stirnhirn. Wien u. Leipzig, 1890.
14. Cunningham, D. J., Contribution to the Surface Anatomy of the Cerebral Hemispheres. Dublin, 1892.
15. Ecker, A., Die Hirnwindungen des Menschen. Braunschweig, 1869.
16. Wernicke, C., Lehrbuch der Gehirnkrankheiten. 1881, pp. 80, 83, and 85.
17. Wernicke, C., Beiträge zur Anatomie des Gehirns. *Arch. f. Anat. u. Physiol., Physiol. Abt.* (1878), p. 591.
18. Sachs, H., Das Hemisphärenmark, Arb. aus d. Psychiat. Klinik Breslau. Leipzig 1892, Vol. I.
19. Dejerine, J., Anatomie des centres nerveux. Paris, 1901, Vol. I.
20. Probst, M., Zur Kenntnis der Grosshirnfaserung, etc. *Sitzungsb. d. (Kaiserliche) Akademie der Wissensch. Math. Naturw.*, CXI-CXII (1902-3), 581.
21. Schnopfhagen, F., Die Entstehung der Windungen des Grosshirns. *Jahrbuch f. Psychiat.*, IX (1890), 197.

22. Vialet, —., Note sur l'existence à la partie inférieure du lobe occipital d'un faisceau d'association distinct, etc. *Comptes rend. Soc. de Biol.*, XLV (1893), 793.
23. Probst, M., Über die Leitungsbahnen des Grosshirns. *Jahrb. f. Psychiat.*, XXIII (1903), p. 18.
24. Probst, M., Ueber den Bau des vollständig balkenlosen Grosshirnes, etc. *Arch. f. Psychiat.*, XXXIV (1901), 709.
25. Anton, G. and Zingerle, H., Bau, Leistung und Erkrankung des menschlichen Stirnhirnes. Graz, 1902.
26. V. Niessl—Mayendorf, E., Vom Fasciculus longitudinalis inferior. *Arch. f. Psychiat.*, XXXVII (1903), 537.
27. Meyer, A., The Connections of the Occipital Lobes and the Present Status of the Cerebral Visual Affections. *Transactions*, Association of American Physicians, XXII (1907), 7.
28. Archambault, La Salle, Inferior Longitudinal Bundle and the Geniculo-Calcarine Fasciculus. *Albany Med. Annals*, XXX (1909), 118.
29. Cushing, H., Distortions of the Visual Fields in Cases of Brain Tumour. *Brain*, XLIV (1921–22), 341.
30. Broca, P., Le grand lobe limbique. Mémoires d'anthropologie, Paris, V (1888), 273.
31. Turner, William, The Convolutions of the Human Cerebrum Topographically Considered. *Edinburgh Med. Jour.*, II (1866), 1105.
32. Pansch, A., Die Furchen und Wülste am Grosshirn des Menschen. Berlin, 1879.
33. Wernicke, C., Das Urwindungssystem des menschlichen Gehirns. *Arch. f. Psychiat. u. Nervenkr.*, VI (1875–76), 298.
34. Eberstaller, O., Oberflächen-Anatomie der Grosshirn-Hemisphären. *Wiener Med. Blätter*, VII (1884), 542.
35. Smith, G. E., A New Topographical Survey of the Human Cerebral Cortex. *J. Anat. and Physiol.*, XLI (Third Series, II) (1906–7), 237.
36. Smith, G. E., Evolution of Man. London, Oxford University Press, 1924.
37. Von Economo, C., Zur Frage des Vorkommens der Affenspalte beim Menschen im Lichte der Cytoarchitektonik. *Zeitschr. f. d. ges. Neurol. u. Psychiat.*, CXXX (1930), 419.
38. Jefferson, G., The Morphology of the Sulcus Interparietalis. *Jour. Anat. and Physiol.*, XLVII (Third Series, Vol. VIII) (1912–13), 365.
39. Mihalkovics, Victor, Entwicklungsgeschichte des Gehirnes. Leipzig, 1877, p. 154.
40. Rüdinger, N., Beiträge zur Anatomie und Embryologie als Festgabe Jacob Henle. Bonn, 1882, p. 186.
41. Broca, P., Note sur la topographie cérébrale. Mémoires, Paris, V, 1888.
42. Cunningham, D. J., The Fissure of Rolando. *Jour. Anat. and Physiol.*, XXV (1890), 1.
43. Sherrington, C. S. and A. S. Grünbaum, Motor Areas of the Anthropoid Brain. *Lancet*, II (1901), 935.

44. Wagner, R., Studien über den Hirnbau der Microcephalen, etc. *Abhand. d. k. Gesell. d. Wissensch. zu Göttingen*, X (1861-62), 75. See table 1, left hemisphere, mark C.
45. Symington, J. and P. T. Crymble, The Central Fissure of the Cerebrum. *Jour. Anat. and Physiol.*, XLVII (1912-13), 321.
46. Cushing, H., A Note upon the Faradic Stimulation of the Postcentral Gyrus in Conscious Patients. *Brain* XXXII (1909-10), 44.
47. Campbell, A. W., Histological Studies of the Localisation of Cerebral Function. Cambridge, 1905.
48. Meynert, Th., Psychiatrie. Wien, 1884, p. 40.
49. Dejerine, J. and A. Thomas. Un Cas de cécité verbale avec agraphie suivi d'autopsie. *Rev. Neurol.*, XII (1904), 655.
50. Von Monakow, C., Gehirnpathologie. Wien, 1897.
51. Minkowski, M., Étude sur les connexions anatomiques des circonvulutions rolandiques, pariétales et frontales. *Schweizer Arch. f. Neur. u. Psychiat.*, XII (1923), 227.
52. Smith, G. E., New Studies on the Infolding of the Visual Cortex and Significance of the Occipital Sulci in the Human Brain. *Jour. Anat. and Physiol.*, XLI (1906), 199.
53. Brodmann, K., Vergleichende Lokalisationslehre der Grosshirnrinde. Leipzig, 1909.
54. Brodmann, K., Beiträge zur histologischen Lokalisation der Grosshirnrinde, Mitteilung I. Die Regio Rolandica. *Jour. f. Psychol. u. Neurol.* II (1903-4), 79.
55. Vogt, O., Zur anatomischen Gliederung des Cortex cerebri. *Jour. f. Psychol. u. Neurol.* II (1903-4), 160.
56. Vogt, O., Furchenbildung und architektonische Rindenfelderung. *Jour. f. Psychol. u. Neurol.*, XXIX (1922-23), 438.
57. Von Economo, C., Die Bedeutung der Hirnwindungen. *Allg. Ztschr. f. Psychiat.*, LXXXIV (1926), 123.
58. Von Economo, C., and G. N. Koskinas, Die Cytoarchitektonik der Hirnrinde des erwachsenen Menschen. Springer, Wien u. Berlin, 1925.
59. Ariëns Kappers, C. U. Signification des fissures du cerveau. *Névraxe*, XIV-XV (1913), 217.
60. Ariëns Kappers, C. U. The Evolution of the Nervous System in Invertebrates, Vertebrates and Man. Haarlem, 1929.
61. Turner, Wm., The Convolutions of the Brain, etc. *Jour. Anat. and Physiol.*, XXV (1890-91), 105.
62. Cunningham, D. J., On Cerebral Anatomy. *British Medical Jour.*, II (1890), 277.
63. Baillarger, quoted by Broca in his essay, Le grand lobe limbique, etc., p. 289. *See* Number 30 above.
64. Jelgersma, G., Das Gehirn ohne Balken, etc. *Neurol. Centralbl.* IX (1890), 162.
65. Pansch, A., Einige Sätze über die Grosshirnfaltungen. *Centralbl. f. d. med. Wissensch.*, XV (1877), 641.

INDEX

Annectants (annectant gyri), 27
  bridges across the fissures, 27
  cuneo-lingual, anterior, 40, 41
  cuneo-lingual, posterior, 38, 41
  dividing completely the central fissure, 79
  gyrus cunei of Ecker, 39, 41
    in the anthropoid brain, 39
  incomplete barrier to the intercortical pathways, 1
  interruptions in intercortical fibers, 1, 27, 115
  projection fibers, relation to, 27
  relation of, to subcortical laminae, 28
  significance of, in central fissure, 70, 71
  significance of, in interparietal fissure, 64
  superior temporal fissure, in course of, 106
  three kinds of, 27
Anterior commissure, 34
Arcuate fasciculus, *see* Intercortical pathways, deep
Areas
  central, 70
  different cell structure of, and connections between, 4, 118
  frontal, 34, 58, 84, 96, 119
    topography of, 84
  insula (Island of Reil), 63, 92, 93, 119
  mesial, 29
  occipital, lateral, 109, 112
  occipital (lobe), 34, 58, 92, 104, 120
  orbital, 89
  parietal, 34, 59, 91, 100, 106
  peristriata, 118
  prefrontal, 34, 84, 89, 104
    topography of, 84
  pyriform, 52
  rhinal, 58
  striata, 40, 118
  temporal, 34
    lateral, 105
  temporo-occipital, 105
    basal, 52
  temporo-parietal, 92
  visual, 40, 58
Capsula externa, 100

INDEX

Casts, plaster of Paris and metal, 11
    drawings from, 12, 13, 14, 15, 17, 31, 48, 50, 53, 60, 75, 78, 81, 94, 97, 101
Cells
    cortical areas, of different cell structures, 4
    damaged by the explosion of the cerebrum, 6, 11
    fusiform, of the lowest cortical stratum, 25
    origin of the intercortical fibers, 25
Cerebrum (human)
    laminated structure of white substance of, 6, 7, 8, 9, 10
    manual separation of laminae of, in exploded preparations, 11
    topography of human, 4
    types of, 4
    water lost in, after explosions, 11
Convolutions (gyri)
    angular, 96
    cingulum, 35, 39, 41, 42, 47, 49, 51
    cuneus, 38, 41, 42, 46, 49, 51
    frontal, middle, 73
    fusiform, 41, 42, 46, 52, 54
    hippocampal, 39, 51, 52, 58
    lingual, 38, 47, 51, 54, 58
    obverse of crests of, 11
    paracentral (lobule), 34, 67, 72
    parietal (lobule), inferior, 46, 100
    parieto-temporal, 96
    pattern of, 4
    postcentral, 72, 73, 91, 96
    precentral, 72, 73, 100
    precuneus, 35, 41, 42, 51
    rectus, 49, 89
    rostral, 58
    supramarginal, 96
    temporal, second, 96
Corona radiata, 47, 100
Corpus callosum, 33, 34, 47, 49, 56, 58, 91
    forceps major, 49
Cortico-spinal tract, 72
Dissection
    automatic, 5
        cell injuries caused by, 6, 11
        quantity of water expelled, 11
        sections of inked preparations, 6, 7, 8, 9, 10
    dissociation of laminae in exploded preparations, 11
    flattened, 16, 19
    manual, 3, 11
    records of, 11. *See also* Casts
Drawings
    detailed microscopic, 20, 48, 76, 80, 82, 93, 98, 99, 101, 103
        method of preparing, 20

INDEX

Drawings, *continued*
  from casts, *see* Casts
  line, 14, 15, 53, 60, 94
  schematic, *see* Schemata
Eminentia collateralis, 52
Fasciculus, *see* Intercortical pathways, deep
Fissuration
  caused by a disproportion of surface and volume, 120
  caused by resistance of skull capsule, 119
  causes of a general pattern of, 2
  causes of a particular pattern of, 2
  contribution to the knowledge of, 117
  effect of the subcortical pathways on, 2
  examples of, in different bodies, 121
Fissures
  boundaries between areas of different structure, 2, 117, 118, 119, 122
  calcarine, 40
  calcarine, anterior, 39, 41, 42, 46, 49, 54, 58
  calcarine and parieto-occipital, 36, 41
    in the ape, 39, 40, 41
    in the human fetus, 40, 41
    morphology of, 39
    three other types of, 44
  calcarine, posterior, 36, 39, 46, 49, 51, 54
    different types accounted for, 36
  central, 72, 87, 88, 116
    annectants, 70, 71
    communication with the Sylvian fissure explained, 71
    general description, 72
    history and morphology, 70
    left, 72
    rare types of, 74
      left, 79
      right, 79
    right, 74
  cingulate, 33, 52, 58
    general considerations of, 57
    gross form of, 34
    late development, explanation of, 34
  collateral, 41, 52, 58
    description of, 52
    general considerations of, 57
    hippocampal portion of, 52, 54, 58
    late appearance, explanation of, 52
    occipital portion of, 52
    temporo-occipital portion of, 42, 52, 58
  cuneus, sulcus of, 36
  frontal
    inferior, 84, 87, 88

Fissures, frontal, *continued*
    middle (Eberstaller), 84, 85, 88, 89
    superior, 84, 85
  fronto-marginal, 89
  fronto-orbital, 89
    frontal or superior, 89
    fronto-orbital proper, 89
    prefrontal, 89
  incisure (parieto-occipital), 42, 65
  interparietal, 59, 63, 117
    Affenspalte (ape-fissure), 59, 62
    postcentral, inferior, 59, 63, 65, 67, 118
    sagittal (or horizontal ramus or arch), 59, 63, 64, 65, 67, 68
    significance of annectants in the course of, 64
    sulcus lunatus, 62, 65
    sulcus occipitalis transversus, 59, 62, 65, 68, 109, 118
    variations of, 63
  limbic, 58
  occipital lateral
    inferior, 108, 111, 113
    superior, 109, 111
  occipitalis (sulcus)
    anterior, 62
    inferior, 62
  obverse of, 11, 14
  order of appearance, 122
  paracingulate, 29
    another type of, 33
  parieto-occipital, 39, 40, 41, 46, 49, 51, 54, 58
  pattern of, 119
  postcentral
    inferior, *see* Fissures, interparietal
    superior, 63, 64, 66
  precentral
    inferior, 72, 73, 87
    superior, 84, 85
  relation of, to intercortical pathways, 117
  rhinal, 49
  subcentral, 71
  subparietal, 35, 58
  supramarginal, 64, 65
  Sylvian (horizontal), 63, 71, 72, 79, 92, 106, 107
    posterior ascending limb of, 64, 67, 68
      diagonal sulcus, 67
      vertical sulcus, 67
  temporal
    middle, 105, 106, 107
    superior, 62, 105, 106, 107

INDEX

Fissures, temporal, *continued*
  temporo-occipital, 109, 111, 113
  temporo-parietal, 62, 66, 67, 106, 107, 109
  trifissural junction, 41, 42
Geniculate eminence, 56
Geniculo-calcarine radiation, 56
Hallucinations, associated visual and olfactory, 58
Intercortical pathways, deep, 115
  arcuate fasciculus, 54, 57, 72, 91, 92, 96, 115
    fasciculus centralis obliquus, 100
    fasciculus centro-parietalis, 100
    fasciculus centro-parieto-occipitalis, 100
    fasciculus longitudinalis superior, 100
    general considerations of, 102
  inferior longitudinal fasciculus, 54, 115
    degeneration of, in lesions of temporal convolutions, 57
  insula, capsule of, 91, 92, 115
  occipital lobe, 49, 51, 115
    fasciculus occipitalis verticalis, 46
    Vialet's tract, 47
Intercortical pathways, underlying the cortex of the fissures
  bundles, arrangement of fibers in, 23, facing 24
  calcarine, anterior, 41
  calcarine and parieto-occipital, 41
    analysis of three types of, 44
  calcarine, posterior, 38
  central
    left, 72
    rare types of
      left, 79
      right, 83
    right, 74
  cingulate
    one piece type, 35
    two piece type, 35
  collateral
    hippocampal portion of, 54
    posterior or occipital portion of, 52
    temporo-occipital, 54
  complication of pattern of, 29, 114
  cuneus, sulcus of, 36
  frontal
    inferior, 88
    middle (of Eberstaller), 89
    superior, 85
  fronto-orbital
    frontal portion, 89
    fronto-orbital portion, 90
    prefrontal portion, 90

Intercortical pathways, *continued*
  general arrangement of, in the subcortical laminae, 23
  general considerations, 114
  gross form of, 6
  in different types of human cerebrum, 4
  incisure (parieto-occipital), 42
  interparietal
    inferior postcentral, 63
    sagittal ramus, 65
    transverse occipital and sulcus lunatus, 65
  length of, 1
  occipital lateral
    inferior, 111
    superior, 109
  opossum's cerebrum, 23, 102
  origin of, 25
  parieto-occipital, 42
  pattern of, 114
    complication of, 29, 114
  posterior ascending limb of the Sylvian
    diagonal portion of, 67
    vertical portion of, 68
  postcentral, superior, 66
  precentral
    inferior, 87, 88
    superior, 87
  proportion of, to projection fibers, 26
  relation to the fissures, 117
  temporal
    middle, 108
    superior, 106
  temporo-occipital, 113
  temporo-parietal, 69
  trifissural junction, 41, 54
  uncinate fasciculus, 57, 58
Laminae, subcortical, *see* Subcortical laminae
Lamina terminalis, 34
Limbic lobe, 33, 57, 58
Methods
  criticism of, 116
  cross-sectioning of tissue, 3
  degeneration, massive, 3, 56
  employed in present study, 5
  manual dissection, 3
  Marchi, 3, 56, 91, 116
  myelination, 3, 56
  Weigert-Pal, 3, 6, 18, 56, 57, 91, 100, 116, 117
Myelination, inapplicability of, 3
Nomenclature employed, 29

INDEX

Olfactory bulb, 34
Olfactory lobe, 58
Operculum
　fronto-parietal, 100
　parieto-temporal, 67
　superior, 73
Opossum's cerebrum, 23, 102, 103
Optic radiation, 49
Projection fibers, 7, 9, 10, 47
　estimate of relative numbers of, 26
　intercortical fibers as collaterals of, 25
　mixed with intercortical fibers, 2, 98, 99, 101, 116
　numbers of, in rodents, 26
　relation to annectant convolutions, 27
Pulvinar, 56
Records, 11, 14.　*See also* Drawings
　casts, 14
Rhinencephalon, 33, 34
Schemata, 30, 32, 37, 43, 55, 61, 77, 86, 95
　method of preparation, 21, 76, 77
Sections, cross, of the cerebrum
　along the course of nerve bundles, 20, Fig. 9 *facing* 20, Fig. 10 *facing* 22, Fig. 11 *facing* 24, 48, 76, 80, Fig. 27 *facing* 80, 82, 93, 98, 99, 101, 103
　authors' complaints regarding, 3, 7, 8, 9, 10
　serial, of exploded preparations, 6, 7, 8, 9, 10
Septum lucidum, 47
Subcortical laminae, 12, 13, 17, 19, 48, 50, 101, 115, 116
　average thickness, 16
　changes caused by annectants, 28
　general arrangement of fibers in, Fig. 9 *facing* 20, Fig. 10 *facing* 22, 23, Fig. 11 *facing* 24, 48, 80, Fig. 27 *facing* 80, 82, 93.　*See also* Schemata
　general shape of, 16, 17
　method of preparing flat blocks from, 16, 19
　mixed nature of fibers, 2, 98, 99, 101, 116
Stratum profundum convexitatis, 91
Tangenitial fiber systems, 2, 122
　of the opossum, 102, 103
Tapetum, 47, 56
Temporo-orbital recess, 89
Thalamus, 56
U-fibers, 1
　in the cerebellum, 2
　length of, 1
　origin of, 25
Ventricle
　lateral, 47
　temporal, 49, 56
　temporo-occipital, 56, 57
Visual radiation, 56, 57

COLUMBIA UNIVERSITY PRESS
COLUMBIA UNIVERSITY
NEW YORK

FOREIGN AGENT
OXFORD UNIVERSITY PRESS
HUMPHREY MILFORD
AMEN HOUSE, LONDON, E.C.

Bei Fragen zur Produktsicherheit wenden Sie sich bitte an:
If you have any questions regarding product safety,
please contact:

Walter de Gruyter GmbH
Genthiner Straße 13
10785 Berlin
productsafety@degruyterbrill.com